Ajay Kumar Goswami

Spectrophotometric Determination of Scandium, Zirconium and Titanium

Also of Interest

Diffusion and Electrophoretic NMR
Peter Stilbs, 2019
ISBN 978-3-11-055152-5, e-ISBN 978-3-11-055153-2

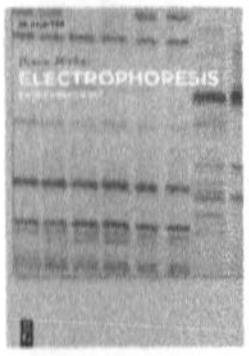

Electrophoresis.
Theory and Practice
Budin Michov, 2020
ISBN 978-3-11-033071-7, e-ISBN 978-3-11-033075-5

Atomic Emission Spectrometry.
AES - Spark, Arc, Laser Excitation
Heinz-Gerd Joosten, Alfred Golloch, Jörg Flock, Susan Killewald, 2020
ISBN 978-3-11-052768-1, e-ISBN 978-3-11-052969-2

Chemical Analysis in Cultural Heritage
Volume 1
Luigia Sabbatini, Inez Dorothé van der Werf (Eds.), 2020
ISBN 978-3-11-045641-7, e-ISBN 978-3-11-045753-7

Ajay Kumar Goswami

Spectrophotometric Determination of Scandium, Zirconium and Titanium

Reagents and Methods

DE GRUYTER

Author
Prof. Ajay Kumar Goswami
Subhash Nagar
132 Road No. 5
Udaipur-313001
India
akumargoswami56@gmail.com

ISBN 978-3-11-162666-6
e-ISBN (PDF) 978-3-11-162683-3
e-ISBN (EPUB) 978-3-11-162685-7

Library of Congress Cataloging-in-Publication Data
A CIP catalog record for this book has been applied for at the Library of Congress.

Bibliographic information published by the Deutsche Nationalbibliothek
The Deutsche Nationalbibliothek lists this publication in the Deutsche Nationalbibliografie; detailed bibliographic data are available on the Internet at http://dnb.dnb.de.

Cover image: onurdongel / iStock / Getty Images Plus
Typesetting: VTeX UAB, Lithuania
Printing and binding: CPI books GmbH, Leck

www.degruyterbrill.com
Questions about General Product Safety Regulation:
productsafety@degruyterbrill.com

Contents

1 Introduction to spectrophotometric determination of Sc/Zr/Ti

It is an established fact of analytical chemistry that spectrophotometric determination methods are one of the most common methods used for transition metals. Dating back to onset of the development of coordination chemistry, the simplest versatile methods were spectrophotometric or colorimetric methods using two components—a metal with orbitals and a complexing agent called an organic reagent—the result being a metal complex or a coordination compound. Such chemical reactions are very important in understanding the reaction kinetics, thermodynamic and kinetic stability of metal complexes and lastly, but perhaps most importantly, the structure and geometry of these complexes. An entire array of possible applications has opened up and helped to develop an entire branch of useful chemistry involving medicine and pharmaceutical chemistry. The principal advantage of these method is that they are highly sensitive, universal in determining both small and large amounts of substance in question and use simple instrumentation, not to mention their cost effectiveness. The increasing number of synthetic organic reagents has further enhanced their application to the effect that almost all metal can be determined successfully using one of the reagents. Furthermore, spectrophotometric instrumentation also has been developed for various typical applications using combinations with other methods and available computer software.

The phenomenal development in synthetic strategies and methods has significantly increased the number of organic reagents for spectrophotometric applications. There is now a list of organic reagents that form complexes with high extinction coefficients, making it possible to reliably determine trace levels of metals. The selectivity and reliability are two factors that make a good organic reagent. Reliability means independence from small changes in reaction conditions. Further, slight differences of other substances present in the solution on the base reaction do not effect. In general when the elements in acidic medium do not need any buffer solution the organic reagent produces reliable results. Thus, the method used in preparing the solution for the spectrophotometric measurement is simple dilution of the analyte with an acid. Moreover, the possible interference associated with the hydrolysis of the ions and the formation of polymeric structure is minimized. This is especially significant in the determination of certain rare and heavy elements like Ti, Zr, Hf etc. The following is a list of reagents proposed in recent years used in highly acidic media:

1. Arsenazo III and analogues
2. Antipyrine and its derivatives
3. Triphenyl methanes
4. Fluorones
5. Pyrogallol compounds
6. Pyridine and other heterocyclic compounds

https://doi.org/10.1515/9783111626833-001

1.1 Importance and use of color reactions for the determination of Sc, Ti and Zr using spectrophotometry

1.1.1 Reagents

Principally, organic reagents and the development of color by metal complexes are important and vital in the spectrophotometric determination of transition metals. This warrants understanding of fundamentals of metal complexes and methods used to determine them. Since this book is focused on methods of determination and reagents used, it is of utmost relevance to discuss and describe in brief the essential parts of both. In the following section, the principal characteristics or rather qualifications of organic reagents are described and the characteristics that make a reagent effective or suitable for determination using complex formation.

1.1.1.1 Contrast

An important characteristic of any color reaction is contrast, which is in often confused with sensitivity, though it is not well-defined in analysis. Any reaction exhibits contrast when two components of the metal complex formation, i. e., metal and ligand (organic reagent or reagent), differ significantly in color under the reaction conditions defined. Quantitatively, the contrast of any reaction is the difference $\Delta\lambda$ in nanometers between the band maxima in the spectra of the reaction product and reagent. It can be expressed as

$$\Delta\lambda = \lambda \text{ Product max} - \lambda \text{ Reagent max.}$$

The ranges defined for poor, moderate and good contrast are

$$\begin{aligned} &\Delta\lambda < 40\,\text{nm}, &&\text{poor,}\\ &\Delta\lambda = 40\text{–}80\,\text{nm}, &&\text{moderate,}\\ &\Delta\lambda > 80\,\text{nm}, &&\text{high contrast.}\end{aligned}$$

There are in practice methods to improve the contrast so as to enhance the application of a given method. When a reagent with –OH group is used, the contrast can be enhanced by use of a solvent that promotes the elimination of a proton from OH group. These solvents include dimethylformamide, dimethyl sulfoxide or any proton-accepting solvent. Another way to enhance or increase contrast of a given color reaction is modulating the structure of a reagent by substitution of a better chromophoric group. Many electronegative groups like NO_2, Cl and Br can do this task. A number of monoazo compounds containing NO_2 are examples of this type.

1.1.1.2 Sensitivity

The quantitative expression of a color reaction is called sensitivity, which may be quantitatively described in terms of molar extinction coefficient ε. The measure in general for the scale of sensitivity is as follows:

When: $\varepsilon < 2 \times 10^4$ (the reaction is relatively insensitive)
$\varepsilon \approx 2\text{–}6 \times 10^4$ (it is moderately sensitive)
$\varepsilon > 6 \times 10^4$ (it is highly sensitive, the limiting value being is 2×10^5), and

when $\varepsilon > 2 \times 10^5$, these are considered to be cascade reactions (in cascade reactions, isolation of intermediates is not required, each reaction compositing the sequence occurs spontaneously, the reaction conditions do not change among the consecutive steps of a cascade reaction and no new reagent are added after the initial steps).

The average molar coefficient is calculated using

$$\varepsilon = A/Cl,$$

where A is the optical density of the analyte; C is the concentration of substance to be determined; and l is the thickness of the solution for which photometric measurements are to be made. The true molar extinction coefficient is a measure of how efficiently a molecule absorbs light at a given wavelength. The sensitivity of any reaction depends on the properties of the organic reagent and on the molar extinction the coefficient of the product. However, reaction conditions such as pH, solvent, temperature etc. have little influence on the sensitivity. By understanding the mechanism and alteration done on reaction can help to increase sensitivity of a chemical reaction.

1.1.1.3 Selectivity

The concept and understanding of both specificity and selectivity have been studied by number of authors [8, 9, 10]. The Selectivity of any organic reagent is the range of reactive components that can react with the particular reagent to give color of the mixture to be analyzed. We can improve the selectivity by variation of pH of the medium, particularly with the reagent of ROH type. In general, to improve selectivity of a method, one of the following methods may be used:

1. Using masking agent or complexones.
2. Determining the elements as a part of mixed ligand complexes.
3. Employing an extraction–spectrometric method.
4. Introducing into the molecule a reagent of substituents that create steric hindrance in the vicinity of chelating center, which is Most versatile method to improve selectivity.
5. Introducing into the reagent molecule a second complex forming group together with the main functional groups. Commonly, synthesis of reagents with heterocyclic rings containing S, P, N and other atoms also improve selectivity.
6. Using different tautomers of a reagent.

Thus, there are numerous ways to enhance selectivity of a reaction for spectrophotometric determination. However, there is no universal method to do this, and it depends on the structure of reagent molecule and nature of metal to be determined.

1.1.2 Methods to increase efficiency of spectrophotometric analysis

It has been made possible to at least increase the efficiency of a method by following three procedures:

- Making multicomponent complexes.
- Altering the reaction medium by applying mixed solvents or even ion aqueous solvents.
- Using surfactants.

These three procedures enable increasing the sensitivity and also shift selectivity of the reaction to the desired direction. A brief description is included in following paragraphs to help readers better understand the methods.

1.1.2.1 Mixed ligand complex formation

It is possible to enhance the characteristics of any color reaction by switching to a more complex reaction from a binary one. This has been incorporated into the spectrophotometric methods for determination. This is because mixed ligand complexes constitute the nearest form in which ions in general exist, although hetero ligand complexes are used more in spectrophotometric methods for extraction. Examples of this category are components of organic cations with metallochromic chelates such as using quaternary ammonium salts with such systems.

1.1.2.2 Organic solvents in spectrophotometry

Employing organic solvents that not only constitute medium but also have significance in the reaction mechanism is one way to improve a procedure and also explore new kinetics and mechanism of a system. This procedure is extremely important as well as having enormous variations. The observed variations are due to the complexity of the process by which complex formation occurs. The solvent can displace tautomeric as well as acid–base equilibria and alter both solvate and reactants, as well as reaction products.

The class of reagent commonly employed are preferably miscible with water, except in cases where solubility is an issue. Alcohols, acetones, dioxin, acids or even DMSO, DMF etc. are some of the common solvents applicable in spectrophotometric determinations. In conclusion, organic solvents are used to improve both sensitivity as well as selectivity of a particular procedure, without having much impact on the mechanistic aspect of color reaction.

1.1.2.3 Application of surfactants

While surfactants in spectrophotometric analysis are not new, they sometimes produce excellent results when applied. Surfactants are amphiphilic organic molecules having one or more hydrocarbon chains (hydrophobic traits) to which is attached a polar group or hydrophilic head. They create self-assembled molecular clusters that are called micelles and absorb onto the interface between a solution and other phase, namely, gases/solids. They are classified into either the ionic or nonionic class. Furthermore, ionic surfactants are of two types: anionic where hydrophilic group dissociates into anions in aqueous solution, and cationic surfactants which dissociates into cations. In addition, amphoteric surfactants can dissociate into anion or cation based on pH conditions. It is important to know that water-soluble ionogenic or noniogenic surfactants are the commonly used ones. We list some of the surfactants of both categories here:

A. *Cationic surfactants*
   (i) Salts of aliphatic primary/secondary/tertiary amines.
   (ii) Quaternary ammonium bases.
   (iii) Alkyl pyridinium/nitrogen containing bases like imidazolines.
   (iv) Imidazoles.
   (v) Pyrimidines.
   (vi) Oxazolines.
B. *Anionic surfactants*
   (i) Alkene sulfonates.
   (ii) Alkylarene sulfonates.
   (iii) Alkyl sulfates.
   (iv) Compounds containing additional polar groups such as carbamide, ester etc. in the chain are categorized as anionic surfactants.
C. *Nonionogenic surfactants*
   The application of nonionogenic surfactants, or the nonionic surfactants that are nondependent on surface-active anions for effect, changes the mechanisms of their action. This occurs both with organic reagent and with complexes formed via reaction. Obviously, the corresponding color is observed only after achieving the critical micelle concentration formed with the surfactant. Application of surfactants along with organic reagents multiplies the analytical use of this class, and multifold new ideas emerge.
D. *Choice of a reagent for individual elements*
   There has been vast development in the field of analytical reagents, in terms of synthesis, suitability and application. Thus, there is an increasing possibility of choosing the right organic reagent for the spectrophotometric determination of an individual element. Author proposes there should be a development of software based on available data to guide an analyst before hand to find such an organic reagent tailored to any individual element. This would definitely help in hastening the analysis and development of effective methods. Presently, there are several thousand organic reagents available. The basic difficulty for an analyst is not in finding a suitable reagent, but is choosing the best among them. A literature survey does not necessarily guide one to a specific reagent that is the best for the given ion or comparatively better than others available. In recent years, literature reviews addressing relevant problems normally are comparative studies of the data characterizing

the efficacy of a given organic reagent for the determination of individual element. To select an organic reagent, the parameters for deciding selection of an organic reagent comprise the following parameters:

1. Sensitivity, based on molar extinction coefficient value.
2. Stability expressed in terms of stability constant.
3. Detection limit of the reagent, based on selectivity and pH.

There have been attempts to classify or make group reagent for both spectrophotometric and extractive spectrophotometric determination of elements. Table 1.1 lists some reagents used frequently for Ti, Zr and Sc.

Table 1.1: Examples of group reagents.

Element	Reagent
Zr, Ti	Chrome Azurol S
Zr, Sc	Anlipyrine C
Sc, Ti, Zr	Arsenazo I
Zr, Sc	Arsenazo III
Zr	Arsenazo M., Dithizone, Xylenol orange, Thoron 1
Ti	Sodium diethyldithiocarbamide xylenol orange
Ti, Zr	Methylthymol blue
Ti, Sc	Pyridylazoresorcinol
Zr, Sc	Sulphochlorophenol S

1.2 Action of organic reagents

Theoretical principles and their development are most important to understand mechanistic aspect of organic reagents. The rules underlying such principles are the following:

(a) Understanding and determining reliable thermodynamic and kinetic characteristics of the processes leading to specific analytical effect.
(b) Understanding mechanism of complex formation, which includes finding consecutive steps in the formation of reaction product as well as elucidating the structure of organic reagent and final reaction product.
(c) Studying effect that determines the nature of color reactions in case of spectrophotometry, luminescence etc.

There have been flaws in the development and understanding of the theoretical, as well as mechanistic, part of structures of complexes with organic reagent. Babko [4] mentioned that even with known organic reagents, the structural formula of complexes has been a matter of tradition, passing from one publication to another, with no justification. This indicates a lag in the development of a general theory of action of organic reagent, as compared to achieving their practical application.

Alimarin [1] studied and developed the concept surrounding the role of functional groups in organic reagents, clarifying the relationship between the structure and properties of the reagent [2]. However, it is beyond the scope of this book to give a detailed review on the state of the art of aspects of the theoretical development and understanding of organic reagents to find more suitable reagents. A characteristic feature of this theoretical development is associated with use of quantum chemical and other such methods to develop more specific and selective reagents for the spectrophotometric determination of elements.

1.3 Spectrophotometric reagents: a brief introduction

In view of the fact that most spectrophotometric methods are based on using organic reagents [32, 6, 35, 38], it is warranted to present introductory details of the various reagents used. The literature shows many reviews published on organic reagents [36, 28, 7]. What's more, series of books on the spectrophotometric methods for metals like Cu, Fe, Ni, Co, V, Cr, Mn, Pd, Pt have been published by the author [14, 12, 11, 13]. This makes it difficult to review the entire scenario of reagents and methods, and thus only a brief introduction of some common reagents is presented here.

1.3.1 Azo reagents

Azo dyes offer sensitive color reactions with metals, and this group includes reagents like 1-(2-pyridylazo)-2-naphthol (PAN), 4-(2-pyridylazo) resorcinol (PAR), 5-Br-PADAP, chlorophosphanazo III and arsenazo III etc. For the determination of metals [3, 29], pyridylazo and thiozolylazo reagents offer important class of reagents.

a. 1-(2-pyridylazo)-2-naphthol (PAN)

PAN can exist in three forms depending on pH of the solution. At pH ≤ 2, it shows a water-soluble yellow-green protonated H_2R^+ form. In the range pH 3–11, it exists as a neutral HR molecule that is soluble in organic solvents to give yellow color. Above pH ≥ 11, it exists as a water-soluble red R^- ion. The structure of the reagent is shown in Figure 1.1.

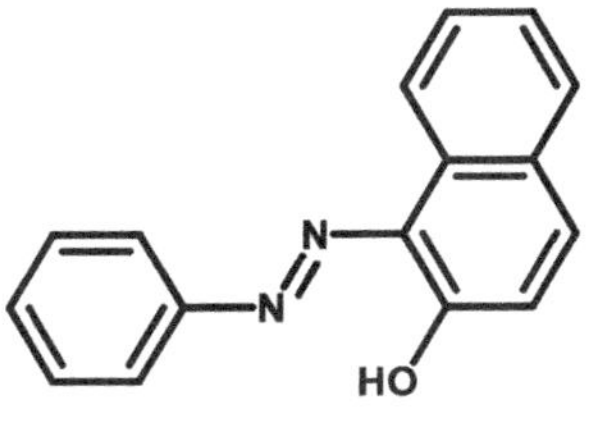

Figure 1.1: 1-(2-pyridylazo)-2-naphthol.

PAN acts as a tertradentate ligand that has donor sites with metal through hydroxyl oxygen, pyridine nitrogen atom and one of the azo nitrogen. Normally, the molar absorptivities of PAN complexes are in the range 2–6 × 10^4. The advantage with PAN is that the absorption maxima λ_{max} of the metal complexes are much different from the reagents, making the determination easier.

b. 4-(2-pyridylazo) resorcinol (PAR)

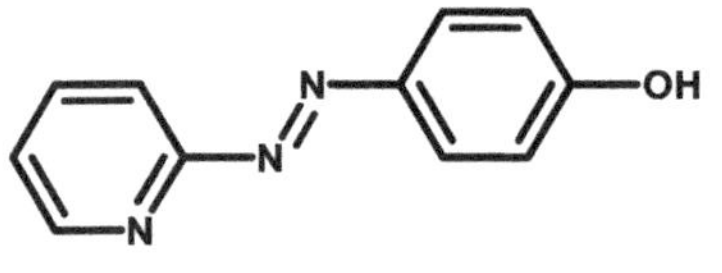

Figure 1.2: 4-(2-pyridylazo) resorcinol.

The reagent is water soluble and forms water-soluble metal complexes so that spectrophotometry can be performed in aqueous medium. In 0.5–0.05 m H_2SO_4, PAR (Figure 1.2) forms complex with Zr, Pd and Ti (III) ions and, on the other hand, in pH 3–6, it produces a color reaction with a variety of ions like Zn, Co, Na, Hg and V, among others. The color of the complexes when using PAR is usually red or violet.

It is of interest that the metal complexes of PAR are extractable into chloroform in the presence of diphenylguanidine [25]. Highly sensitive methods using bromo and chloro derivatives of PAR are also significantly important. The one that is most widely used is 2-(5-bromo-2-pyridylazo)-5-dimethylaminophenol (5-BrPADAP), which has become useful reagent for Cd, Co and Ni, among others. The structures of such reagents can be expressed in Figure 1.3:

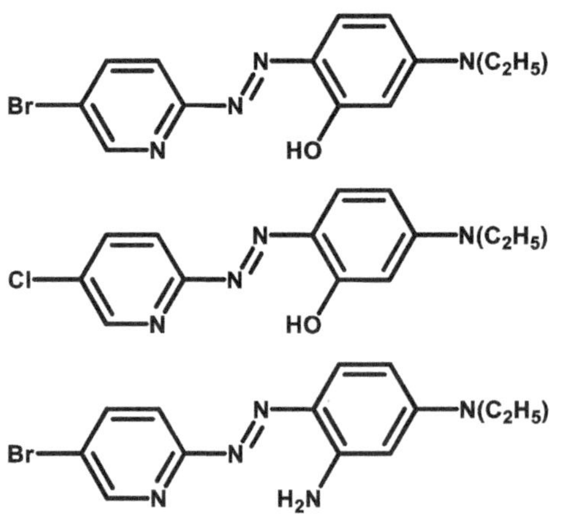

Figure 1.3: Derivatives of 4-(2-pyridylazo) resorcinol.

c. Thiazole azo compounds

These compounds have similar properties as PAN and PAR for application as spectrophotometric reagents. While their stability is lower, their selectivity as reagents offers a large advantage. Some examples are shown here:

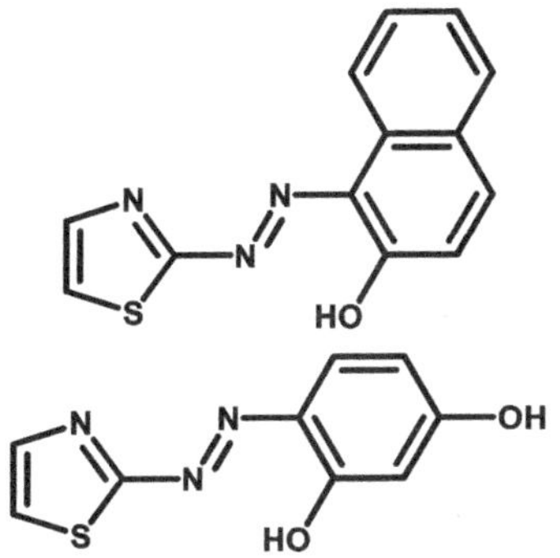

Figure 1.4: Thiazole azo.

A review on thiazylazo reagents (Figure 1.4) and their application as spectrophotometric reagent was published [16].

d. Arsonic azo compounds

Arsenazo (I) (Figure 1.5) and thoron (II) are the first reagents of this group that are characterized by the presence of an arsenic acid group at an ortho position to the azo group. The –OH group also is ortho to the azo group. Structures can be shown as

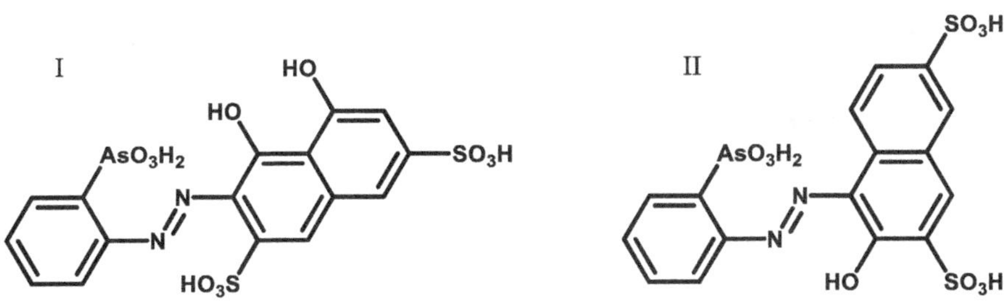

Figure 1.5: Arsonic azo compounds.

Stable complexes are formed due to the presence of ASO_3H_2 even in fairly acidic medium, azo groups being a good donor that forms complex easily and the OH group further forms a second ring with the metal. Not only are the complexes formed stabilized but the obtained color is also improved. However, the presence of sulphonic group helps in making the complex water soluble, making the entire measurement easier. It should be mentioned that hardly any sensitive reagent was available until arsenazo I was introduced, particularly for Th, Zr, Hf, and U, plus rare earths.

It is relevant here to stipulate that arsenazo I and its analogues are derivatives of chromotropic acid or 1,8-dihydroxy napthalene-3,6-disulphonic acid. It is applied as a reagent for the determination of rare earths, Th, U, Zr, Ti, Nb etc., whereas thoron I is comparatively less sensitive and is most commonly used as a reagent for the determination of Li, U, Zr, Be and Th.

e. Arsenazo III

This is a very good bis-azo dye which is based on chromotropic acid and o-amino phenylarsonic acid. It is readily soluble in alkaline solutions as well as moderately soluble in neutral and acidic solutions. The reagent decomposes with strong oxidizing agent like H_2O_2, Cl_2 and Br_2, while reducing agents like $TiCl_3$ also cause decomposition. Arsenazo III (Figure 1.6) has a purplish-red color form at 10m HCl to pH 4 and, at higher pH values, attains a blue-violet color. The structure is represented in Figure 1.6.

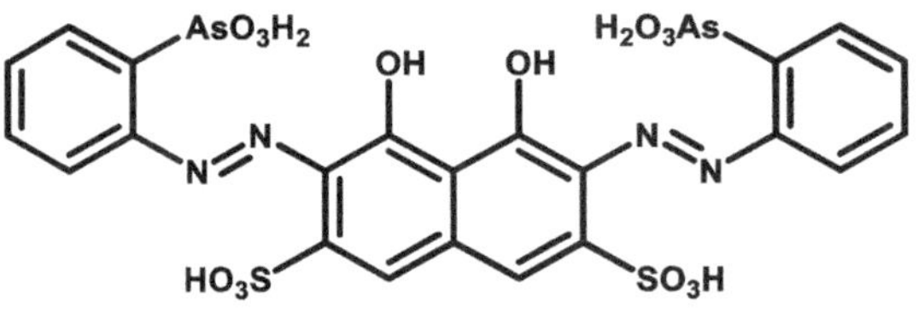

Figure 1.6: Arsenazo III.

The reagent reacts in strong acidic medium 1–10 m HCl, for example, with Th, Zr, Hf and U (IV), with a good molar-absorptivity value is 10^5. On the other hand, the reagent reacts with U (VI), Sc, Fe (III) and rate earth in the pH range 1–4. In such conditions, the sensitivity of this color reaction reduces to $\varepsilon = 5 \times 10^4$. Thus, to avoid difficulties with hydrolysis of some ions like Zr, Th and U, a strongly acid medium can be recommended, which also enhances selectivity with the reagent.

The reagent absorbs at λ_{max} 520–530 nm, but the metal complexes have a slightly higher λ_{max} of 655–665 nm. If we look into the mechanistic side of arsenazo III, only one side of the symmetrical molecule is associated in the formation of complexes with the metal. The metal coordinates through the nitrogen atom of the azo group, the oxygen atom of the arsonic group and the oxygen atom of the hydroxyl group. This causes distortion of the symmetry of the molecule, giving two neighboring absorption maxima in the visible region for the arsenazo III–metal complexes. In concentrated H_2SO_4, it absorbs at 675 nm and can be determined. Other similar reagents are azo dyes containing phosphonic acid groups. An example is chlorophosphanazo III (Figure 1.7), which is useful for the determination of Zr, Ti, U, Sc, Ca and Sr [34].

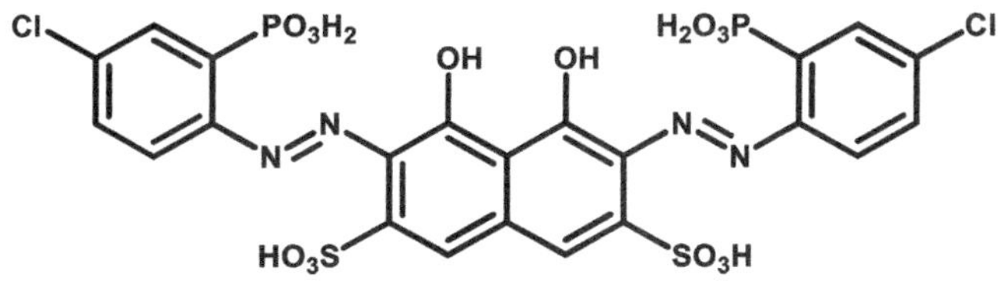

Figure 1.7: Chlorophosphanazo III.

f. Miscellaneous azo compounds

There are number of similar azo compounds applied to the spectrophotometric determination of metals like Nb, Zr, Mo, Cu, V, Al etc. in acid medium, structure is depicted in Figures 1.8–1.11.

(i) Sulphochlorophenol S

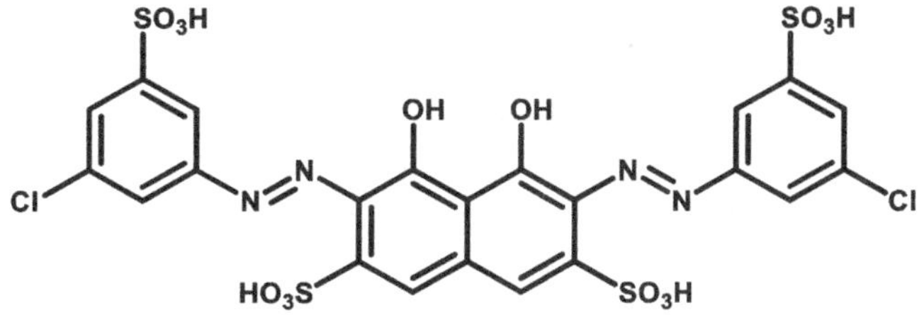

Figure 1.8: Sulphochlorophenol S.

(ii) Sulphonitro phenol M

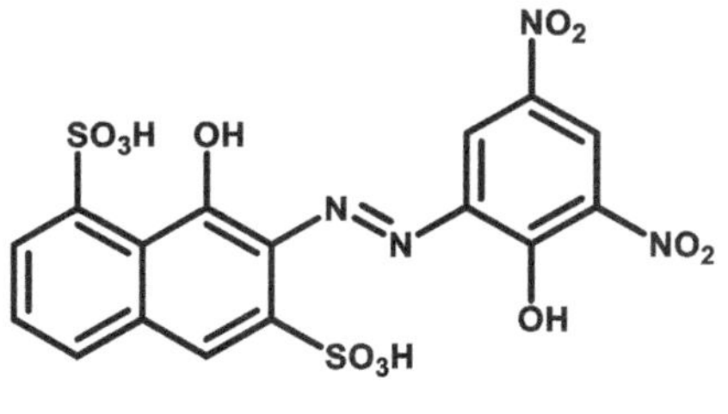

Figure 1.9: Sulphonitro phenol M.

(iii) Chromotrope 2B

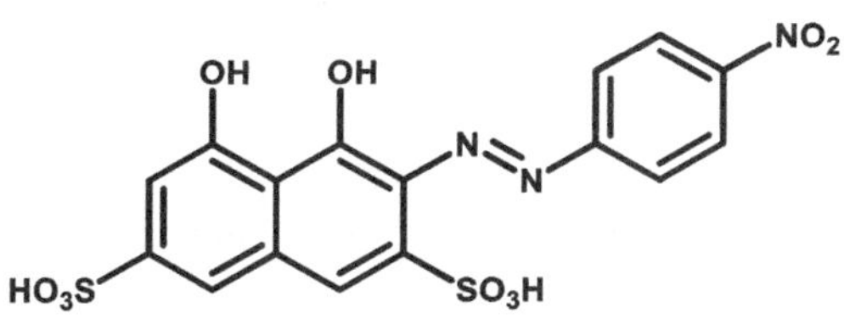

Figure 1.10: Chromotrope 2B.

(iv) 2-(4, sulphophenylazo chromotropic acid (SPADNS)

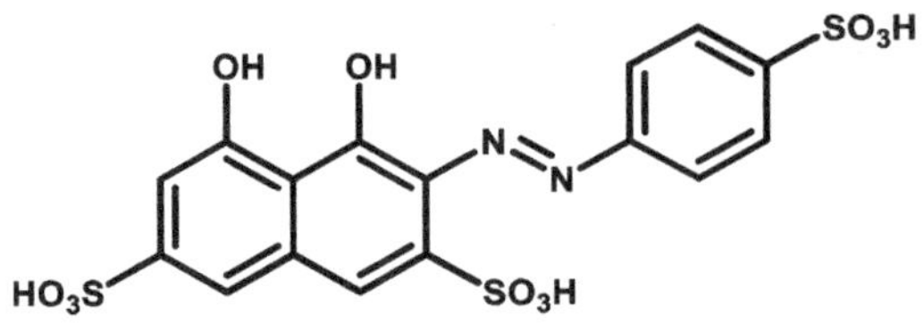

Figure 1.11: 2-(4, sulphophenylazo chromotropic acid.

Excellent review on 8-hydroxyquinoline azo derivatives, pyrocatechol azo derivatives and o-thiazo derivatives of p-cresol and 2-naphthol [20, 33, 30] are available for the more keen reader.

g. Triphenylmethane reagents

The presence of p-quinonoid rings in triphenylmethane class of reagents make them intensely colored and thus make them extremely good reagents for the spectrophotometry of number of metals. This is a vast area, and details can be extensive so a few examples are provided to make the reader aware of this class.

(i) **Pyrocatechol violet**

It is a commonly used chelating reagent that is soluble and aqueous ethanol. In the pH 1–8 range, the aqueous solution of reagent is yellow in color, which changes to violet with measuring pH. Almost all chelates formed with metals with the reagent are blue, under weakly acidic to weakly basic conditions. Figure 1.12 shows the structure of the reagent.

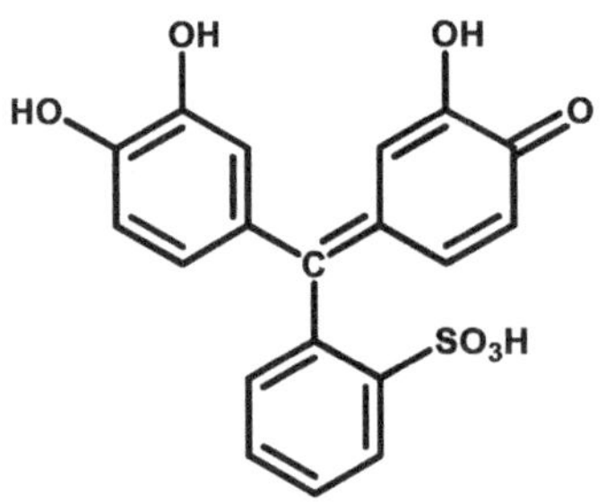

Figure 1.12: Pyrocatechol violet.

(ii) **Eriochrome Cyanin R (ECR)**

This reagent has additionally a carboxylate group besides hydroxyl and sulphonate groups. In weakly acidic solution, ECR reacts with many metals. The original yellow-orange color of the reagent turns to violet or blue as this forms complexes with any metal (Figure 1.13).

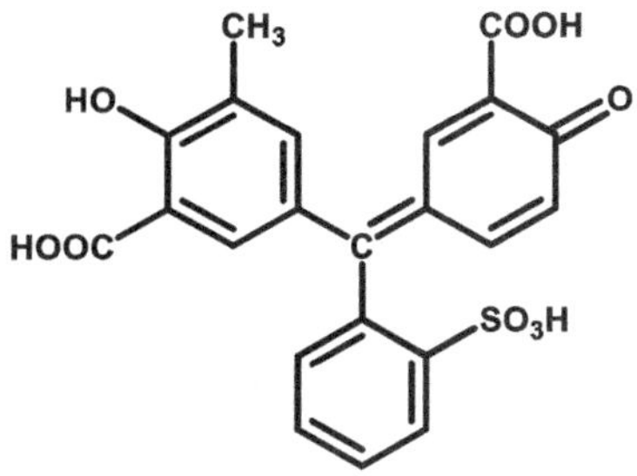

Figure 1.13: Eriochrome Cyanin R.

(iii) **Chrome Azurol S (CAS)**

The reagent CAS reacts with many metal in weakly acidic or neutral medium. It also changes to blue or violet when forms complex with metal. Figure 1.14 represents the structure of CAS.

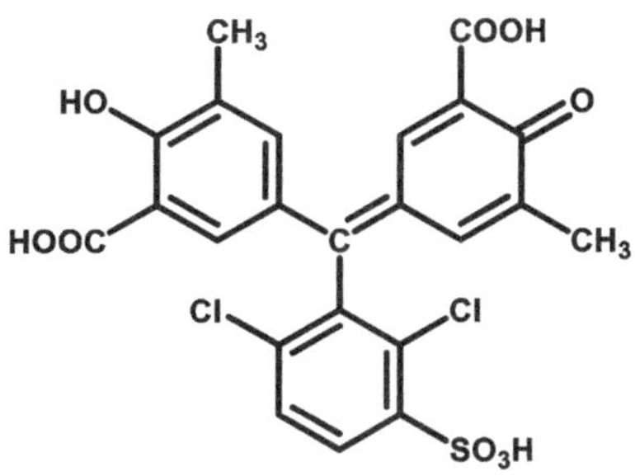

Figure 1.14: Chrome Azurol S.

(iv) **Xylenol orange**

The presence of iminodiacetic acid group characterizes xylenol orange and its similar compound methylthymol blue. It has a similarity with EDTA. Both xylenol orange and methylthymol blue are good chromogenic reagents for Sc, Ti and Zr since these are easily hydrolysable. The values of ε exceeds 10^5. The structures and represented in Figures 1.14 and 1.15.

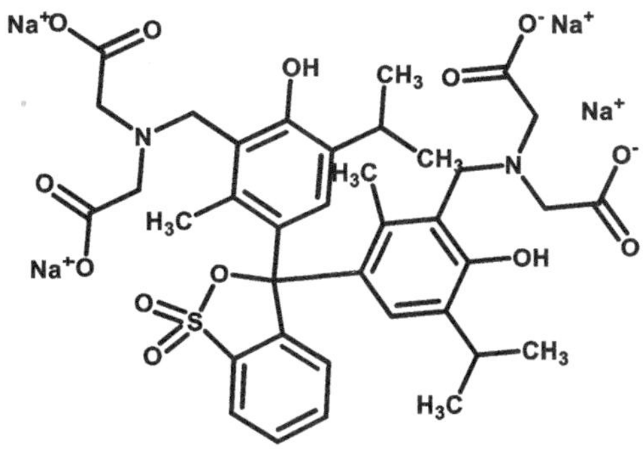

Figure 1.15: Xylenol orange.

h. Xanthene reagents

A number of xanthenes reagents are available. yet most commonly used are:
(1) Gallelin.
(2) Bromopyrogallol red.
(3) Pyrogallol red.

They are used for the spectrophotometric determination of Ti and Zr as well as Bi, In, Mo, Sn and Sb, and Th, among others. In case a cationic surfactant is used, the sensitivity of such determination is enhanced. The structures of these three reagents are represented in Figures 1.16–1.18.

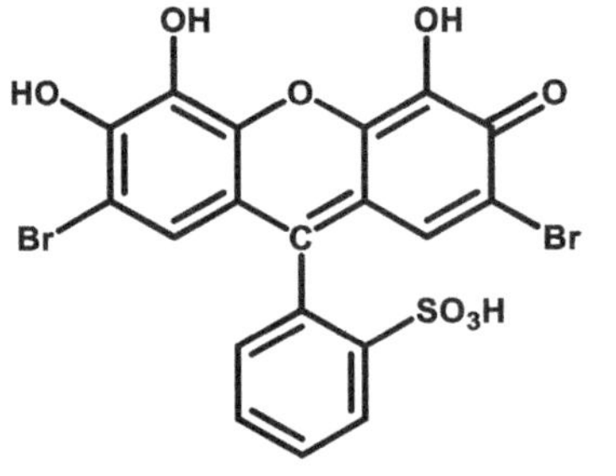

Figure 1.16: Bromopyrogallol red.

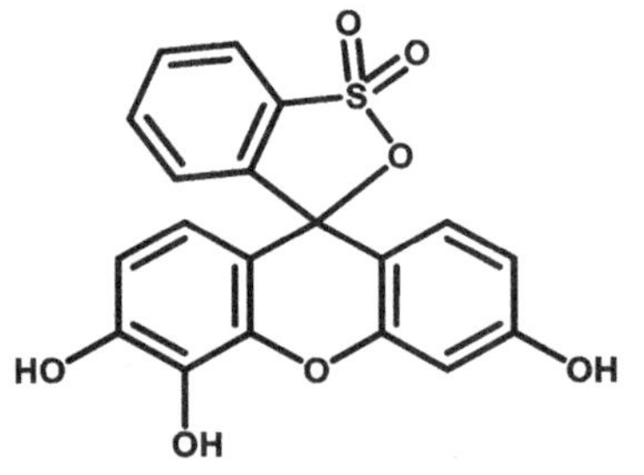

Figure 1.17: Pyrogallol red.

Calcein, also called fluorexone and fluorescein complexone, is a commonly used reagent for calcium and has a similar structure as shown in Figure 1.18.

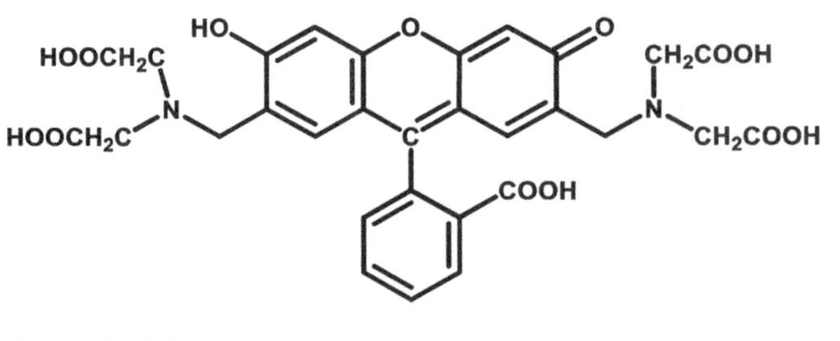

Figure 1.18: Calcein.

1.4 Non-chelating organic reagents

(i) **Basic dyes**

These dyes are commonly used as halide salts. A number of elements form anionic complexes with such reagents. They many form ion pairs with basic dyes that are extractable into nonpolar solvents. Thus, these extracts are very useful for sensitive extractive spectrophotometric methods [39, 26]. The form of dye applied is an ion-associate or singly charged cation. The solvents that are used include benzene, toluene, chloroform or 1,2-dichloroform ethane. To improve the extraction method, a small amount of a donor-active oxygen containing solvents like MiBK, butanol or even di-isopropyl ether may also be used. Moderately concentrated mineral acids such as HCl, H_2SO_4 etc. are used in the pH range 3–5 to control the acidity of the aqueous phase.

(ii) **Triaryl methane dyes**

A commonly used dye is triphenyl methane, which is included in class of TAM dyes. The group is represented by brilliant green and, malachite green. Their structures are represented in Figures 1.19–1.20.

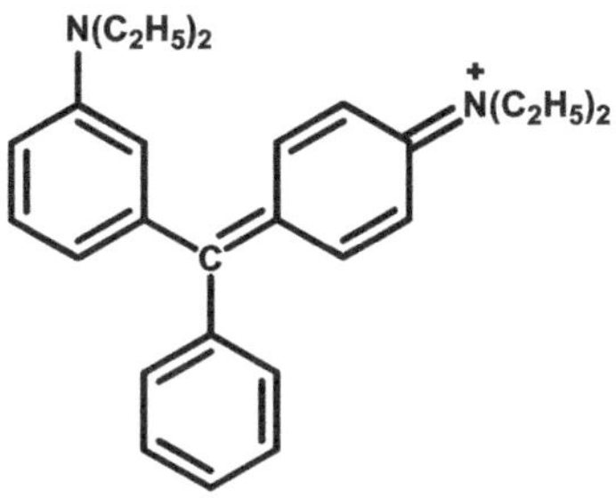

Figure 1.19: Brilliant green.

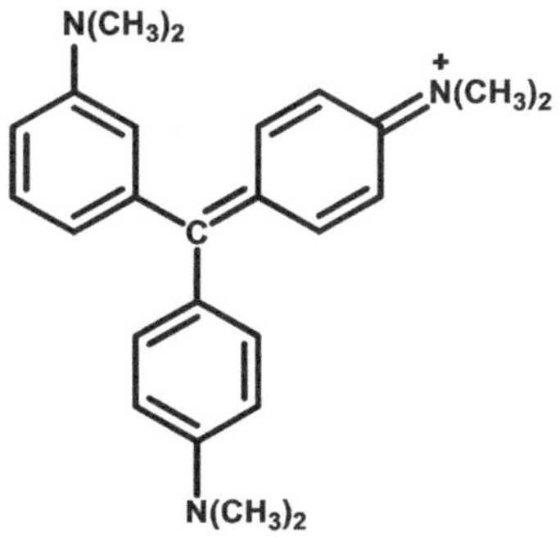

Figure 1.20: Malachite green.

The other dyes included in this class are rhodamine 3B, butyl rhodamine B and rhodamine 6G, which are xanthene basic dyes.

1.5 Acid dyes

In general, acid dyes are less commonly used compared to basic dyes. They form extractable ion-associates with hydrophobic cationic metal complexes. They commonly known as acid-base indicators (Figures 1.21–1.23). The acid xanthenes dyes include erythrosine, rose bengal B and eosin. They are applied as sodium salts to facilitate solubility issues. The structures shown next are representative of this category.

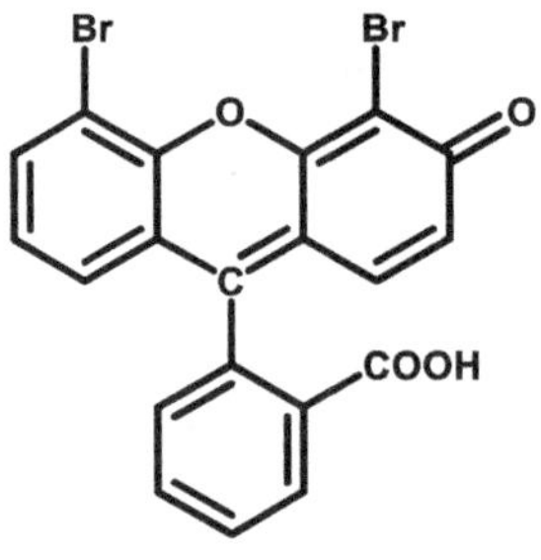

Figure 1.21: Eosin.

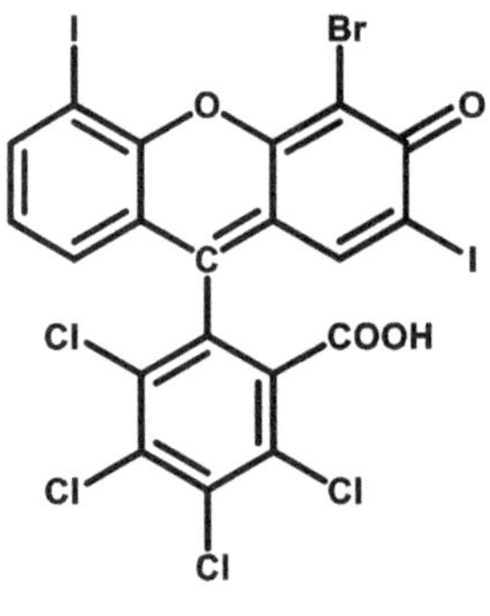

Figure 1.22: Erythrosin/rose bengal B.

Another dye in this class is bromophenol blue.

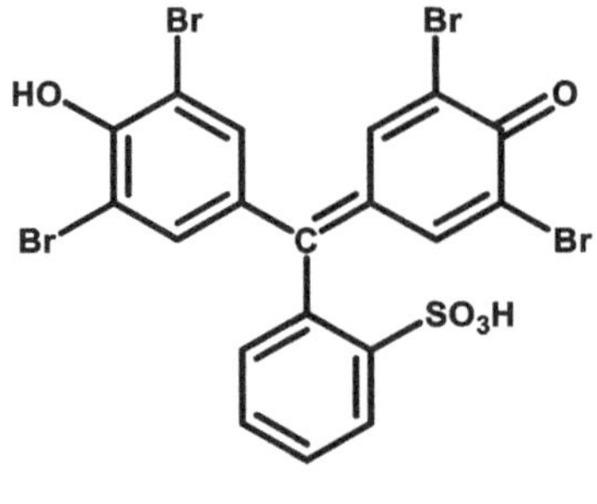

Figure 1.23: Bromophenol blue.

1.6 Dithizone

The most common and frequently used reagent is Diphenylthiocarbazone, 3-mercapto-1,5-diphenyl-formazane, called Dithizone (H_2DZ). It is used in extractive separation and the spectrophotometric determination of a number of metals [21, 19, 18]. The reagent is insoluble in aqueous solutions at < 7, though it dissolves in basic media. It dissolves in CCl_4, $CHCl_3$ and hydrocarbons as well as alcohols giving a green solution. The structure is shown in Figure 1.24.

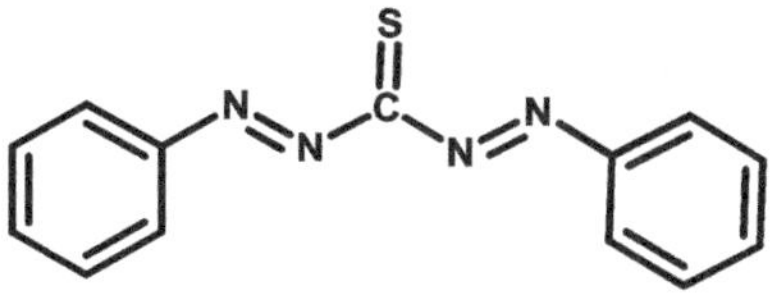

Figure 1.24: Dithizone.

The reagent binds metal through sulfur atoms by replacing hydrogen in the thiol group. The other side is through the nitrogen atom [23].

Another analogue of this reagent is Di-(2-naphthyl) thiocarbazone (dinaphthizone), which is also a reagent for spectrophotometry, although the metal complexes of this reagent are less soluble compared to dithizone complexes.

1.7 Dithiocarbamates

A common reagent of this category is sodium diethyldithio carbamate (Na-DDTC) for spectrophotometry [17, 15, 5].

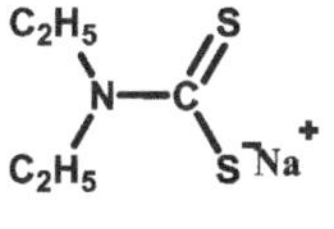

Figure 1.25: Dithiocarbamate.

The metal complexes of dithiocarbamates (Figure 1.25) are poorly soluble in water. They dissolve in organic solvents like $CHCl_3$, CCl_4, acetone etc. Figure 1.26 depicts the structure of diethylammonium diethyldithiocarbamate.

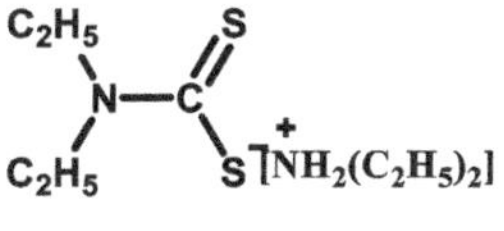

Figure 1.26: Diethylammonium diethyldithiocarbamate.

The class of analytical reagent similar to DDTC diethylammonium is stable in acid solution and dissolves in $CHCl_3$. Common analogues of this category are dibenzyl dithiocarbamate, pyrazoline dithiocarbamate and pyrrolidine dithiocarbamate [24].

1.8 8-hydroxyquinoline (oxine)

The reagent 8-hydroxyquinoline (I), also called oxine, is a very common spectrophotometric reagent. The nature of the reagent is amphoteric, and it is soluble in $CHCl_3$, C_6H_6, CCl_4 and many other organic solvents. The distribution between solvent and water depends on the pH matrix of the solution. Other derivatives of oxine are chloro-oxine (II) and bromo-oxine (III). The structures are shown in Figure 1.27.

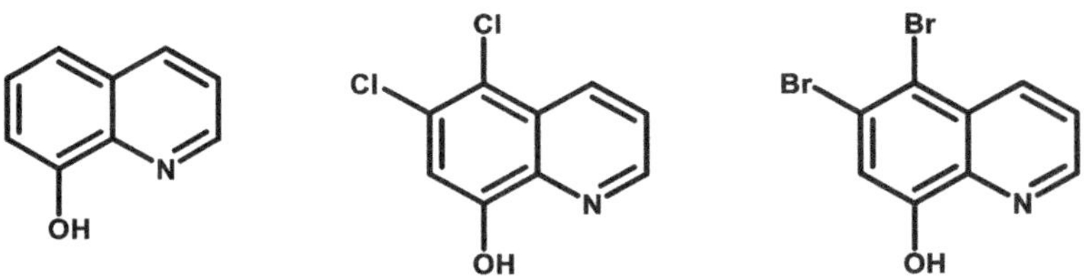

I II III

Figure 1.27: Oxines.

Other similar reagents are ferron and 8-mercaptoquinoline, which are slightly soluble in chloroform. The molar absorptivities of thio-oxine and oxines vary from 5×10^3 to 1×10^4. There is an exhaustive list of other reagents used for spectrophotometric determination of metals that is beyond the scope of present book. The application for each metal in question, i. e., Sc, Zr and Ti will be presented in subsequent chapters. However, to make a cursory mention, a list of such reagents is being given here.

(a) **Hydroxylamine derivatives:** The reagent is used for determination of many metals including Ti, Nb, Fe (iii) etc.
(b) **Hydrazones:** A triatomic =C=N–N≡ is characteristic functional group of hydrazones pyridime-2-aldehyde-2-quinolylhydrazone (PADQH) and cuprizone are examples of this category of reagent.
(c) **Dioximes:** These reagents are represented by dimethylglyoxime and α-furyldioxime.
(d) **β-diketone:** Acetylacetone and thienoyltrifluoro acetone are examples of this category of spectrophotometric reagents for metal determination. Fluorinated β-diketones have been used for the determination of lanthanides [37].
(e) **Thio-Michler's Ketone:** This is an important and useful reagent for the spectrophotometric determination of a number of metals.
(f) **Crown-ethers/cryptands:** They are also useful as reagents for the spectrophotometric determination of a number of metals [27].

1.9 Inorganic reagents

An important inorganic reagent is thiocyanate, which is most commonly used in the spectrophotometric determination of numerous metals like Fe (iii), Mo, W, Nb and Ti [31, 22].

Other inorganic reagents include heteropolyacids, hydrogen peroxide, $SnCl_3^-$ ion and chloride and azide.

The present chapter gives an overall view of various reagents used, but, while one can still include many more, it is not possible to include them due to the limited scope of this book which is intended address specifically the methods.

1.10 Theoretical prediction of organic reagents for inorganic analysis using computational methods

Of late, artificial intelligence has become a buzzword in all academic and related fields. The problems hitherto being solved were especially difficult in chemistry as well as other fields. Last many years computational drug designing has become an area which addresses excellent software based prediction, molecular modeling, prediction of activity structure correlations, docking to select target based drugs, etc. successfully. Selection of a suitable structure-based analytical reagent for a particular metal, a-priori can be a challenge. To be precise, there are hardly any such tools to predict choosing a good reagent for the spectrophotometric determination of as metals. References that show individual methods and reagents can be an exhaustive task, and thus the published data is helter-skelter. At this stage, very few references are available on this topic, and the situation needs to be addressed soon. To conclude this chapter and the recommended state-of-the art strategies, this section is a preliminary attempt to touch upon this unresolved area of interest. Spectrophotometric determination is based on two fundamental components:

A. Reagent
B. Method

Since the UV-visible method is entirely based on color reactions between metals and ligands (reagents), the entire gambit becomes the reagent chosen. If, by any means instead of searching thousands of molecules available, an AI based tool can predict the best selective and sensitive reagent for a particular metal, it would save both time and energy for the method-development people. So, there are current attempts to explore this part, namely, the theoretical prediction of an organic reagent for inorganic analysis, using computational chemistry. The principle is to identify molecules with specific structural and electronic properties that allow them to selectively chelate or bond with the target metal ion. This involves analyzing factors such as the molecular donor atoms or sites, charge distribution, the potential for complex formation or stability of the complex formed, including thermodynamic and kinetic parameters, on this basis, to predict a suitable reagent for the analytical procedure.

1.11 Fundamental aspects of theoretical prediction

a. **Molecular modeling**
   This is done by using software program to generate a 3D model of potential organic molecules and simulate its interaction with a particular metal ion. This should enable evaluation of factors like bond lengths, coordination geometry and stability of the complex formed.
b. **Quantum-chemical calculations**
   Methods like DFT (density functional theory) may be used to calculate the critical electronic properties of a molecule. The factors like electron density distribution HOMO/LUMO energies and charge distribution are very important to understand metal–ligand interactions.
c. **Ligand design principles**
   These are facilitated on the basis of desired selectivity for a particular metal ion, choosing functional groups that have proper donor atoms such as oxygen, nitrogen, sulfur etc. that form strong bonds with the metal under consideration.
d. **Stability constant calculations**
   This is done for predicting the stability constants of the metal complexes formed by a metal with the specific ligand on the basis of thermodynamic parameters, to identify potential interferences from other metal ions.
e. **Molecular docking simulations**
   This is done to identify potential binding sites on the metal ion and find out the probable organic molecules' coordinating geometry.
f. **Chelating ability**
   This predicts on the basis of donor atoms the capability for forming chelating rings with the metal ion.
g. **Selectivity**
   This has to be found out to predict that the organic reagent could ideally react preferentially with the chosen metal ion in the presence of other present ions.
h. **Sensitivity**
   It is the parameter that explains that complexes formed should have detectable properties like vibrant color, change or fluorescence so that detection can be done easily.
i. **Solubility**
   This factor determines that a chosen reagent should be soluble in a particular solvent, ensuring good mixing and the homogenous nature of the matrix.

Finally, experimental validation of the theoretical predictions made on the basis of the cited data and advanced computational methods and machine learning algorithms can be utilized to analyze large databases of organic reagents and metal complexes to predict new candidates with the desired properties.

1.12 Spectrophotometry—a versatile tool for analysis

Color in a compound is explained by the criteria that the compound absorbs, transmits or reflects light over an electromagnetic spectrum in the visible region, when light passes through any such solution of the compound. The reflected part is partially a fraction of the light that it absorbed. This characteristic reflected light the in visible region is used as a measurement of the concentration of the chemical solution of the metal complex. Spectrophotometry is an important method for studying the metal complexes of transition-metals. UV-visible spectrophotometry has thus become an indispensable part of analytical chemistry, more so of metal-complex studies. Since this book focuses on the spectrophotometric determination of Sc, Zr and Ti, a complete subsequent chapter is dedicated to address these details.

Bibliography

[1] Alimarin, I. P., Frid, B. I.: New method for detecting small quantities of columbium and tantalum in minerals, rocks, and metals. *Zavod. Lab.* 6, 823 (1937)

[2] Alimarin, I. P., Burova, T. A.: Quantitative separation of columbium and tantlum with sodium hypophosphite. *Zh. Prikl. Khim.* 18, 289–293 (1945)

[3] Anderson, R. G., Nickless, G.: Heterocyclic azo dyestuffs in analytical chemistry. A review. *Analyst* 92, 207 (1967)

[4] Babko, A. K., Pilipenko, A. T.: *Fotometricheskii Analiz (Photometric Analysis).* Izd. Khimiya, Moscow (1968)

[5] Bayo, S.: Calculation of the optimal experimental conditions for liquid—liquid extractions with diethyldithiocarbamic acid. *Anal. Chim. Acta.* 105, 281 (1979)

[6] Cheng, K. L., Ueno, K., Imamura, T.: *Handbook of Organic Analytical Reagents.* CRS Press, Boca Raton (1982)

[7] Dalawat, D. S., Chauhan, R. S., Goswami, A. K.: Review of spectrophotometric methods for determination of zirconium. *Rev. Anal Chem.* 24(2), 75–102 (2005)

[8] Feigle, F., Anger, V.: *Spot Tests in Inorganic Analysis.* 6. Elsevier, Amsterdam (1972)

[9] Fiegle, F.: *Quantitative Analysis by Spot Tests: Inorganic and Organic Applications*, 2nd English edn. Elsevier, Amsterdam (1939)

[10] Fiegle, F.: *Chemistry of Specific, Selective and Sensitive Reactions.* Academic Press, New York (1949)

[11] Goswami, A. K.: *Spectrophotometric Determination of Palladium and Platinum Methods and Reagents.* CRC Press, Boca Raton (2023)

[12] Goswami, A. K.: *Spectrophotometric Determination of Nickel and Cobalt Methods and Reagents.* De Gruyter GmbH, Berlin/Boston (2024)

[13] Goswami, A. K.: *Spectrophotometric Determination of Vanadium, Chromium and Manganese Methods and Reagents.* De Gruyter GmbH, Berlin/Boston (2024)

[14] Goswami, A. K., Agarwal, S.: *Spectrophotometric Determination of Copper and Iron Methods and Reagents.* De Gruyter GmbH, Berlin/Boston (2021)

[15] Halls, D. J.: The properties of dithiocarbamates: a review. *Mikrochim. Acta* 57, 62–77 (1969)
[16] Hovind, H. R.: Thiazolylazo dyes and their applications in analytical chemistry. A review. *Analyst* 100, 769 (1975)
[17] Hulanicki, A.: Complexation reactions of dithiocarbamates. *Talanta* 14, 1371 (1967)
[18] Irving, H. M. N. H., Iwantscheff, G.: The analytical applications of dithizone. *Crit. Rev. Anal. Chem.* 8, 321–366 (1980)
[19] Irving, N. M.: *Dithizone*. Royal Society of Chemistry, London (1977)
[20] Ivanov, V. M., Rudometkina, T. F.: *Zh. Anal. Khim.* 33, 2446 (1978)
[21] Iwantscheff, G.: *Das Dithizone Lind Seme Auwendung in der Mikro-Lind Spine analyze*, 2nd edn. Verlag Chemie, Weinheim (1972)
[22] Juriaame, A., Kemp, D. M.: *Talanta* 15, 1287 (1968)
[23] Lang, M., Alsop, P. A.: *Talanta* 17, 242 (1970)
[24] Likussar, W., Boltz, D. F.: Spectrophotometric determination of extraction constants for certain metal 1-pyrrolidinecarbodithioates. *Anal. Chem.* 43(10), 1273–1277 (1971)
[25] Mapuliya, S. G., et al.: *Zh. Anal. Khim.* 35, 1307 (1980)
[26] Marczenko, Z.: Nichtchelatierende Farbstoffe als Basis empfindlicher spektrophotometrischer Methoden zur Bestimmung von Elementen. *Mikrochim. Acta II* 68, 651–684 (1977)
[27] Mchedlov-Petrosyan, N. O., et al.: Ultraviolet and light absorption spectrometry. *Zh. Anal. Khim.* 49, 177 (1994)
[28] Nawaz-Diaz, A.: *Talanta* 92, 207 (1991)
[29] Pilipenko, A. T., Savransky, L. I.: Microprocessor-controlled system for automatic acquisition of potentiometric data and their non-linear least squares analysis. *Talanta* 25(8), 451–459 (1978)
[30] Pringle, D. L., et al.: Source of error in trace-element analysis of biological samples. *Talanta* 29, 973–984 (1982)
[31] Rózyeki, C.: *Chem. Anal. (Warsaw)* 11, 447 (1966); 14, 755 (1969); 15, 3 (1970)
[32] Sandell, E. B., Onishi, H.: *Photometric Determination of Traces of Metals, General Aspects*. Wiley, New York (1978). (Chapters 5 and 6)
[33] Savos'kina, L. N., Nemodrunk, A. A., Oskotskaya, E. P.: *Zh. Anal. Khim.* 36, 1483 (1981)
[34] Savvin, S. B.: *Organic Reagents of Arsenazo III Group*. Atomizdat, Moscow (1971)
[35] Sommer, L., Ackermann, G., Burins, D. T., Savvin, S. B.: Present and future status of organic analytical reagents—Part I: General remarks. *Pure Appl. Chem* 62(11), 2147–2166 (1990)
[36] Stephen, W. L.: *Analyst* 102, 793 (1977)
[37] Topilova, Z. M., et al.: *Zh. Neorg. Khim.* 34, 2224 (1989)
[38] Townshend, A., Burns, D. T., Guibault, G. G., Lobinski, R., Marczenkoz, N. E. J., Onishi, H.: *Dictionary of Analytical Reagents*. Chapman and Hall, London (1993)
[39] Zadorozhnaya, E. M., Nabivanets, B. I., Masley, N. N.: *Zh. Anal. Khim.* 29, 993 (1974)

2 Spectrophotometric determination of metal complexes

A method to determine by knowing how much light a chemical substance absorbs by measuring the intensity of light in form of a beam passing through a sample solution is called spectrophotometry. Since each compound absorbs or transmits light over a particular range of wave length, the method is extremely important and applicable. This technique has become the backbone for studying metal complexes since they absorb in visible or ultraviolet range and commonly have colored solutions. Quantitative analysis of metal-complexes has seen development in leaps and bounds during recent years, and the technique is one of the most versatile tools to measure reaction kinetics, thermodynamics and the stabilities of almost all metal complexes. Since the methodology of spectrophotometric determination is simple, the economy is good and precision is excellent, the method has become an indispensable evitable tool for analytical chemistry. In the following sections, to convey the conceptual underpinning, a very simple introduction is presented to the principles and application of spectrophotometry.

2.1 Introduction

When a compound absorbs transmits or reflects light over an electromagnetic spectrum in visible region and when the light passes through any solution, a part of its is absorbed. Spectrophotometry is an economic and standard method to measure the concentration of the chemical in the solution. The measurement is in terms of absorbance (*A*) or transmittance (*T*). Thus, any application dealing with chemical substance can use this technique, including biochemistry, material science, chemistry, physics, biology and clinical sciences. Depending on the range of wavelengths of light source, it is classified into two different types:
- UV-visible spectrophotometry, which uses light over 185–400 nm (UV) and 400–700 nm (visible).
- IR spectrophotometry uses light over 700–15,000 nm of electromagnetic radiation.

When a sample solution absorbs light in visible region when it absorbs all the light it is black, and in case it transmits all it is white and on transmitting various other parts and absorbing some of it in visible region it gives out that particular color. For example, if it absorbs red light (~ 700 nm), it appears green which is the complementary color of red. In visible spectrophotometry, we use a prism to filter or narrow down certain wavelength, so as to pass the other particular beam of light through the solution. A simple basic arrangement of a spectrophotometric setup is given in Figure 2.1 to better convey how it works.

https://doi.org/10.1515/9783111626833-002

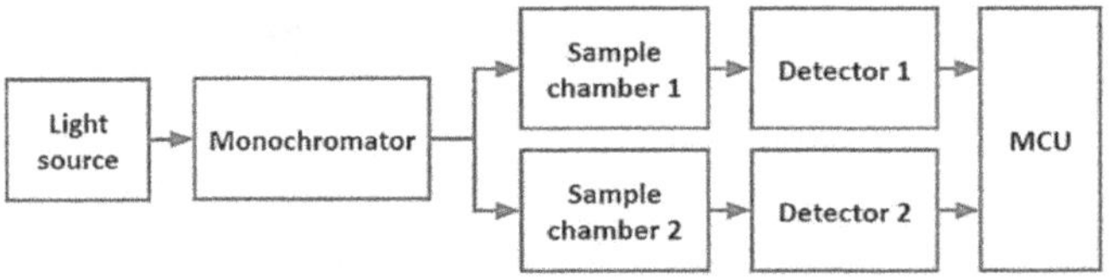

Figure 2.1: Simple setup as a line diagram of a spectrophotometric device.

The spectrophotometer is a combination of two devices, i. e., a spectrometer and a photometer. The former is the part that produces, disperses and measures light, whereas photometer detects or measures the intensity of light.

We use a spectrometer to produce the various wavelengths, as we know compounds absorb best at various different wavelengths. The number of photons that passes through the cuvette is dependent on the length of the cuvette and the concentration of the sample solution. The general relationship between the incident and transmitted beams of light is

$$\text{Transmittance } (T) = \frac{I_t}{I_o},$$

where I_t is the intensity of the transmitted beam of light and I_o is the intensity of incident light. Absorbance is related to transmittance by the following relationship:

$$\text{Absorbance } A = -\log(T) = -\log\left(\frac{I_t}{I_o}\right).$$

2.2 Beer–Lambert law

This is relationship between the amounts of light absorbed by the substance and its concentration. The relationship is expressed by the equation

$$A = \varepsilon CL,$$

where, A is the absorbance of light of specific wavelength, ε is the molar-extinction coefficient, C is the molar concentration of the sample, L is the length of optical path.

When an incident beam of light with intensity I_o passes through a part or fraction of photons, it is absorbed, and the exiting beam has a lower intensity I_t. The Beer–Lambert law applies when L is the thickness of the solution and α is called absorption coefficient. As earlier defined, 'absorbance' is the reverse of transmittance. and it is expressed by

$$A = \log_{10} \frac{I_o}{I_t},$$

and

$$A = \varepsilon CL,$$

where L is the length of the cuvette.

Generally, in UV and visible spectrophotometry, we use 10–390 nm and 390–790 nm as the wavelengths of the visible region. The Beer–Lambert law is the basis of spectrophotometry.

2.3 Methodology of spectrophotometric determination

Although detailed methodology of measurement is available in all standard analytical chemistry textbooks, a general outline is detailed in this section to make the reader aware of the procedure. However, this section is a cursory and precise explanation of the steps used in such methods for metal complexes, in general. The following basic steps comprise the spectrophotometric method.

(a) Choice of wavelength

The selection of a suitable wavelength is the first and foremost step for absorbance measurements. The relevant wavelength of absorbance is done in a region where molar absorptivity (ε) changes quickly with wavelength. Thus, a region of wavelength or particular wavelength is chosen where absorbance is maximum and constant with a small change in wavelength of a colored solution. It is known as λ_{max} or the working wavelength.

(b) Color development

The molecular constitutions of constituents are closely related to the color they exhibit. Thus, as described earlier, the concentration of the constituent is measured by determining the extent of absorption of light of the appropriate wavelength in the colored solution. In the case of metal complexes, sometimes this development of color many not be instantaneous, and it takes some time to reach maximum color intensity. Thus, this has to be taken in consideration when studying metal complexes using spectrophotometric methods.

(c) Deviation from Beer's law

Beer's law is the principle governing spectrophotometric determination. It is expressed as a plot of absorbance versus concentration and is a straight line passing through the origin. It covers a wide range of concentrations in solution state. The color obtained is not normally interfered with by the presence of colored chemical constituents other than the analyte. However, large amounts of such components many cause a shift in maximum absorption. In such conditions, Beer's law is not followed and the system deviates. Thus basically, the systems that strictly follows Beer's law are those that are amenable to determination.

(d) Sensitivity of any spectrophotometric method

The senility of a method is described in terms of the molar absorptivity (ε L/mol/cm) of the metal complex. Regarding suitability of the method for trace determination, sensitivity is a key factor. The expression is

$$\varepsilon = \frac{A}{C.1}$$

The factor on which sensitivity depends is the bandwidth of the monochromatic radiation. In general, the scale for sensitivity is as follows:

$$\begin{aligned} \text{Low sensitivity:} &\quad \varepsilon < 2 \times 10^4 \text{ (L/mol/cm)} \\ \text{Moderate sensitivity:} &\quad \varepsilon = 2\text{–}6 \times 10^4 \text{ (L/mol/cm)} \\ \text{High sensitivity:} &\quad \varepsilon > 6 \times 10^4 \text{ (L/mol/cm)}. \end{aligned}$$

It is also known as the specific absorptivity or Sandell's sensitivity. In fact, it is the mass of analyte per unit volume of the solution. Sandell's sensitivity is the concentration of the analyte (1 µg/ml), which gives an absorbance of 0.001 in a cell path with length of 1 cm, in $\mu g/cm^2$ units.

(e) Precision and accuracy

These two terms are buzz words in analysis of any component using analytical methods. The precision of any measurement is defined as the reproducibility of the set of retrieved experimental data; the accuracy denotes the closeness of a result to its true value. The factors that control accuracy and precision are:

1. Instrumental errors.
2. Chemical variables.
3. Operator's skills.
4. Concentration of an analyte.

(f) Stability constants of a complex

The stability of a metal complex formed determines the selectivity of such a method. The stability constants thus play an important role. The stabilities of chelates depend on a number of factors like the size of the chelate ring, basic strength of the chelating molecule and the nature of the donor atom.

(g) Stoichiometry of the complex

To determine the empirical formula of a complex spectrophotometric method, use one or other of the following:

(i) Job's method of continuous variation.
(ii) Slope–ratio method.
(iii) Turner Anderson's method.

(h) Interfering species

A method is normally more applicable when it exhibits the least interference from other common anions or cations. To minimize this, tolerance limits of the interfering ions are established at those particular concentrations, where there is not more than ± 2.0 % change in absorbance during determination of the analyte under question.

(i) The standard addition method

An ideal calibration standard is made to determine the appropriate composition of any analyte. This is with respect to both concentrations of analyte as well as other species present in the sample. This minimizes the effects of various components in the spectrophotometric measurements.

3 Methods for spectrophotometric determination of scandium (Sc)

3.1 Introduction

The element with an atomic number of 21 is represented by the symbol Sc and found in very small amounts in nature. Since it is mined from the rare minerals in Scandinavia, it is given the name Scandium. The abundance of scandium is the 50th most abundant element on earth and is distributed in more than 800 minerals with a total production at about 50 kg/year. The metal is obtained by heating scandium fluoride with calcium to give the pure metal. The element has application in many types of equipment and glasses, such as lamps and televisions, among others. During the welding of high-strength aluminium alloys, it is used to reduce solidification and cracking. Aluminum–scandium alloys are used in aerospace industry or and even in sports equipment such as baseball and bicycle, to name a couple. When exposed to air, it changes to a yellowish or pinkish color. It dissolves in acids and reacts with hydrogen gas. The metal belongs to the transition metallic group, and due to its limited availability and high costs, it has limited application. Due to a high melting point and low density, it's useful as an alloying agent to impact high performance as in military applications. However, if absorbed by people, the metal concentrates in the liver can contribute health problems; even, some of its compounds are carcinogenic, though the metal itself is not toxic. Although a rare occurrence, it has entered into the food chain in trace amounts. A human can tolerate less than 0.1 microgram per day. It is mostly dangerous due to the fact that gases and damps impregnated with scandium can be inhaled with air. It damages the cell membranes of aquatic animals causing impact on their reproduction and nervous system. Since both negative and positive aspects warrant its determination in a variety of samples, the next section describes methods of its spectrophotometric determination.

3.1.1 Various spectrophotometric methods for the determination of scandium

This section briefly details the spectroscopic methods, e. g., pH, molar absorptivity, Sandell's sensitivity etc. to provide the reader with an adequate picture of the procedures. The methods are ordered chronologically sequence so that oldest to latest methods are treated in one comprehensive view, reagents included. Table 3.1 summarizes them all.

3.1.2 Arsenazo

Hiroshi Onishi et al. [9] used arsenazo for the spectrophotometric determination of scandium in the range of 10–50 µg. The relevant solution with pH 6.1 has an absorbance at 570 nm. It is described that scandium is separated from magnesium, calcium, other

https://doi.org/10.1515/9783111626833-003

Table 3.1: Common analytical reagents used for spectrophotometric determination of scandium.

	Name of the reagent	λ_{max} (nm/pH/solvent)	Molar absorptivity (L/mol/cm)	Sandell's sensitivity (ng/cm^2)
1.	Arsenazo	570 6.1	–	–
2.	Chromazurols	550	27000	.0017
3.	Tiron	310 6.0	8×10^3	–
4.	Bromopyrogallol red	610	2.4×10^4	–
5.	Anthrarufin-2,6-disulphonic acid (disodium salt)	–	–	–
6.	Nalidixic acid	640 4.5	–	–
7.	p-Acetylchlorophosphanazo		2.29×10^5	–
8.	o-Hydroxy hydroquinonaphthalein	– 4.0	1.0×10^5	0.0005
9.	Semi xylenol orange	530	2.4×10^4	–
10.	Indoferron	800	960	–
11.	Eriochromcyanin		1.8×10^5	–
12.	XAD-4 resin impregnated with nalidixic acid	640 4.5	–	1.4 µg/L 0.32 µg/L
13.	Chlorosulphophenol C	649 2.0	7.78×10^5	–
14.	Hexaacetate calyx (6) arene	675 1 M nitric acid (5 pH)	–	
15.	Eriochrome cyanin R (ECR)	610 6.5/(1.0 M HCl)	5.6×10^5 1.0×10^7 M (DL)	–
16.	Nile blue and poly (vinyl alcohol)	585	3.59×10^5	–
17.	2-(5-bromo-2-pyridylazo)-5-diethyl-aminophenol	572 5.3–6.5	7.2×10^4	0.62
18.	N-phenylbenzo-hydroxamic acid (PBHA)	540 4.5	1.7×10^4	
19.	o-Hydroxyhydroquinonaphthaleinin cationic micellar media	5.4	1.0×10^5	0.0005 µg/cm^2
20.	Pyridylazo and thiazolylazo naphthols	545 570	1.9×10^4 to 2.93×10^4	–

rare earths, zirconium, fluoride and phosphate by extraction with TTA in xylene. Furthermore, copper, aluminium and iron (iii) are also removed with 8-quinolinol-chloroform extraction. Even uranium (vi) and thorium are separated using anion exchange. The method thus successfully determines scandium from a complex mixture, as claimed.

3.1.3 Chromazurol S

Ryoei Ishida and Norio Hasegawa [12] developed a spectrophotometric method for the determination of scandium using chromazurol-S in a weakly acidic medium. The λ_{max} of the complex is 550 in the concentration range between 4.27 µg/10 mol. The pH with best color development is at 5.6. The molar extinction coefficient reported is 27,000 with a Sandell's sensitivity of 0.0017 $\mu g/cm^2$. It is further reported that many ions, such as fluoride, phosphate, oxalate, tartaric citrate and EDTA, interfere. Moreover, cations like Al (iii), Be (ii) U (vi), Fe (ii), Th (iv), Cu (ii), Y (iii), Fe (iii), Zr (iv) and W (vi) interfere in the determination as mentioned by the authors.

3.1.4 Tiron

Hiroshi Hamaguchi et al. [4] reported a method for the spectrophotometric determination of scandium with Tiron. It is described that 3–250 µg of scandium can be determined with Tiron. A 1:3 and 1:1 complex of the scandium with Tiron is suggested with the pH 6.0 in acetate buffer. The reported value of extinction coefficient is 8×10^3 at 310 nm. The authors report interference of small amounts of aluminium, titanium, iron, cerium molybdenum, thorium and zirconium. Further, interference of oxalate, citrate, tartarate and malate is also reported.

3.1.5 Bromopyrogallol red

Buneo Shimizu [15] reported a method for the spectrophotometric determination of scandium using bromopyrogallol red. The author reported that this method as highly sensitive as it can which detect a microgram scandium. The report mentioned that at 610 nm and in the pH range of 6–1, scandium can be determined successfully. A 1:1 complex at 610 with molar absorptivity of 2.4×10^4 is reported. Although common cations interfere, complete separation is achievable through successive ion-exchange. It is mentioned that with scandium in traces can be determined within several igneous rocks.

3.1.6 Anthrarufin-2,6-disulforic acid (disodium salt)

John C. Macdonald and John H. Yoe [7] developed a method for spectrophotometric determination of scandium using anthrarufin-2,6-disulforic acid (disodium salt) as an analytical reagent. Up to 2-ppm of scandium can be determined using this method. It is mentioned that 0.04 µg Sc can be determined with sensitivity using this method. To remove interference, a separation procedure has also been developed.

3.1.7 Nalidixic acid

Shabnam Shahida et al. [13] reported an online preconcentration-based spectrophotometric method for the determination of scandium using nalidixic acid as an analytical reagent. The method describes that scandium from aqueous sample solutions of pH 4.5 is selectivity retained in a mini column containing XAD-4 resin impregnated with nalidixic acid, with a flow rate of 1.18 mL/min, in the form of scandium nalidixic acid complex. This complex is desorbed using arsenazo-iii solution and finally measured at 640 nm. The detection limit of the scandium is 1.4–0.32 μ/L. It was mentioned that tolerances from cations such as Th (iv), U (vi) are some rare earths as well as some anions have been improvised. The method as mentioned has been validated on certified materials IAEA-SL-1 (Lake sediment), and the results for scandium are in good agreement.

3.1.8 p-Acetylchlorophosphanazo (CPAPA)

Chung-Gin Hsu et al. [5] developed a spectrophotometric method for determination of scandium based on complex formation of a scandium–cerium–p-acetylchlorophosphanazo system. It was mentioned that a heteronuclear ternary β-complex is formed. The scandium gives a molar absorptivity $\varepsilon = 2.29 \times 102$ l/mol/cm even in the presence of many interfering ions. The method has been extended to scandium in alloys with good results.

3.1.9 o-Hydroxy hydroquinonephthalein in cationic micellar media

Itsuo Mori et al. [8] reported a spectrophotometric method for have determination of scandium (iii) by using ternary complex with o-hydroxyhydroquinonephthalein (Qnph) and N-hexadecyl-pyridinium chloride (HPC) at pH 4.0 as a cationic surfactant. It was determined at 555 nm, with molar absorptivity of 1.0×10^5 dm^3 mol^{-1} cm^{-1} and a Sandell's sensitivity of 0.0005 $\mu g/cm^2$.

3.1.10 Semi-xylenol orange

M. A. H. Hafez et al. [3] developed a method for spectrophotometric determination of scandium (iii) using semi-xylenol orange. In a medium of chloroacetate buffer, the complex scandium–semi-xylenol orange is formed which is measured at 530 nm. The reported molar absorptivity is 2.4×10^4 L/mole/cm.

3.1.11 Indoferron

Tsuneo Shimizu et al. [16] introduced a novel spectrophotometric method for microgram amounts of scandium using indoferron as a reagent. It was measured at 600 nm and has a molar absorptivity of 960 L/mol/mm. The method is free from interference of uranium and other rare earths. It was reported that scandium is separated by a three-stage, ion-exchange chromatographic procedure.

3.1.12 Eriochrome cyanin

Syndorova, Larysa et al. [18] reported a spectrophotometric determination method for Sc (iii) using eriochrome cyanin as a reagent. The effect of cationic water-soluble polyelectrolyte PSPMG on the complex has been studied. The method involves formation of the complex at pH 5.5–6.5 when scandium, ECR and PSPMG in the buffer solution are added which results in a 1:4:1 complex. The reported molar absorptivity value was 1.8×10^5, and the limit of detection was 10^{-7} mol/L of Sc.

3.1.13 Chlorosulphophenol C

Deng, Haixia et al. [2] developed a spectrophotometric method for scandium determination in an HCl-leaching solution of red mud. It was mentioned that reagent chlorosulphophenol C and Sc (iii) form a 1:1 complex at pH 2.0 at 649 nm which can be measured. The reported molar absorptivity value was 7.78×10^5 L/mol/cm and the method can be applied to the determination of scandium (iii) in HCl acid-leaching solution of red mud.

3.1.14 Hexa acetate calix (6) arene

Swamidoss et al. [17] described a method for the extraction-based spectrophotometric determination of scandium (iii) with hexacetate calyx (6) arene. It was detailed that solvent extraction of Sc (iii) was carried out with hexacetate calyx (6) arene in toluene, and Sc (iii) was quantitatively extracted with 1×10^{-4} M of ligand at pH 5 in toluene. The time of equilibration was 12 minutes. Subsequently, the organic phase was stripped with 1M nitric acid to determine Sc (iii) at 675 with arsenazo (iii). The report extends the method in real samples successfully.

3.1.15 Amberlite IRC 718 resin

Park, Chan-Il et al. [10] proposed a spectrophotometric method for scandium with eriochrome cyanin R (ECR). At pH 6.5, 5 ml of 1.0×10^{-2} M ECR, 10 ml of 2.0×10^{-4} M CTMABA form a ternary complex with Sc (iii). This can be measured at 610 nm with a detection limit of 1.0×10^{-7} M. The method was applied to the determination of Sc (iii) in monazite.

3.1.16 Molybdate and Nile blue

Li, Zubi et al. [6] developed a sensitive spectrophotometric method for the determination of trace amounts of scandium using spectrophotometry. The method is based on the formation of ion associate between scandium (iii), molybdate and Nile blue in presence of polyvinyl alcohol. The measurement was done at 585 nm, and the reported molar absorptivity was 3.59×10^5 $dm^3\,mol^{-1}\,cm^{-1}$. The Beer's law linear range is 0–2.0 μg scandium/25 ml. The advantage reported for the method is that tenfold amounts of rare earths do not interfere the determination except for cerium. The method was successful even for geological samples for scandium (iii).

3.1.17 2-(5-bromo-2-pyridylazo)-5-diethylaminophenol

Rathaiah, G. V. et al. [11] published a method for scandium (iii) with 2-(5-bromo-2-pyridylazo)-5-dithylaminophenol as a chromogenic reagent. A 1:1 red-colored complex in the pH range 5.3–6.5 was reported, which is water insoluble but dissolves in $>1 = 32\,\%$ (v/v) methanol. It was measured at 572 nm. The reported molar absorptivity was 7.3×10^4 $dm^3\,mol^{-1}\,cm^{-1}$, and the Sandell's sensitivity value was $0.62\,ng/cm^2$. Additionally, a coefficient of variation of ± 0.43 % with no interference of 50 ions was reported. The described method can be used for the determination of Sc (iii) in Wolframite.

3.1.18 N-phenylbenzohydroxamic acid

Agrawal, Y. K. et al. [1] proposed a method of scandium determination using N-phenylbenzo-hydroxamic acid as a reagent. It was described that scandium (iii) is extracted with N-phenylbenzohydroxamic acid into isoamylalcohol at pH 4.5. A further mixed complex is formed with xylenol orange (XO) or morin. The SC-PBHA–XO/morin complex, a red-colored one, is measured at 540 nm in the Beer's law range of 0.05–3.0 ppm. It had a molar absorptivity and a value of 1.7×10^4 $dm^3\,mol^{-1}\,cm^{-1}$ at 540 nm. It has also a greenish-yellow fluorescence so that scandium can be determined fluorometrically too in range of 0–1.4 ppm as mentioned.

3.1.19 (Pyridylazo)- and (thiazoylazo) naphthols

Sharma, Chander Dhar et al. [14] developed a rapid spectrophotometric method for determining scandium (iii) with the reagent 1-(2-pyridylazo)-2-naphthol (PAN) and 1-(2-

thiazoylazo)-2-naphthol (TAN) in weakly acidic media. The complexes formed are 1:1 and 1:2 chelates with the reagents. The PAN and TAN complexes are reportedly determined at 545 and 570 nm in the Beer's law ranges of 0.16–1.40 and 0.08–1.40 scandium (iii), respectively. The molar absorptivity values are 1.90×10^4 and 2.93×10^4 $dm^3\ mol^{-1}\ cm^{-1}$. The effect of diverse ions on the determination has also been reported.

Bibliography

[1] Agrawal, Y. K., Nagar, A. K.: Solvent extraction and spectrophotometric determination of scandium (iii) with N-phenylbenzohydroxamic acid. *Indian J. Chem, Sect. A, Inorg. Phys. Theor. Anal.* 25(11), 1065–1066 (1986)

[2] Deng, H., Wang, K., Wang, W., Li, A., Zhang, F., Chen, J.: Spectrophotometric determination of scandium in hydrochloric acid leaching solution of red mud. *Chin. J. Rare Met.* 35(5), 764–768 (2011)

[3] Hafez, M. A. H., Kenaway, I. M. M., Kabil, M. A.: Spectrophotometric investigation and determination of scandium using semi-xylenol orange. *Anal. Lett.* 18(16), 2043–2067 (1985)

[4] Hamaguchi, H., Onuma, N., Kuroda, R., Sugisita, R.: Ultraviolet spectrophotometric determination of scandium with tiron. *Talanta* 9(7), 563–571 (1962)

[5] Hsu, C.-G., Liu, S.-C., Pan, J.-M.: Spectrophotometric determination of scandium based on the coloration effect in scandium-cerium-p-acetyl chlorophosphanazo (CPAPA) system. *Talanta* 42(12), 1905–1911 (1995)

[6] Li, Z., Wang, J., Xu, Q.: Spectrophotometric determination of trace amounts of scandium with molybdate, Nile blue and polyvinyl alcohol. *Anal. Sci.* 12(2), 259–262 (1996)

[7] Macdonald, C., Yoe, J. H.: Spectrophotometric determination of scandium with anthrarufin-2,6-disulfonic acid (disodium salt). *Anal. Chim. Acta* 28, 264–270 (1963)

[8] Mori, I., Fujita, Y., Fujita, K., Usami, A., Kawbe, H., Koshiyama, Y., Tanaka, T.: Spectrophotometric determination of scandium (iii) with o-hydroxyhydroquinonaphthalein in catioinic micellar media. *Bull. Chem. Soc. Jpn.* 59(5), 1623–1625 (1986)

[9] Onishi, H., Banks, C. V.: Spectrophotometric determination of scandium with arsenazo. *Anal. Chim. Acta* 29, 240–248 (1963)

[10] Park, C.-I., Cha, K.-W.: Spectrophotometric determination of scandium-(iii) in monazite after separation using amberlite IRC 718 chelating resin. *Bull. Korean Chem. Soc.* 20(12), 1409–1412 (1999)

[11] Rathaiah, G. V., Eshwar, M. C.: Spectrophotometric determination of Sc (iii) with 2-(5-bromo-2-pyridylazo)-5-diethylaminophenol. *Indian J. Chem, Sect. A, Inorg. Phys. Theor. Anal.* 25(1), 101–102 (1986)

[12] Ryoei, I., Norio, H.: The spectrophotometric of scandium (iii) with chromazurol S. *Bull. Chem. Soc. Jpn.* 40(5), 1153–1158 (1967)

[13] Shahida, S., Ali, A., Haleem Khan, M.: On-line spectrophotometric determination of scandium after preconcentration on XAD-4 resin impregnated with nalidixic acid. *J. Iran. Chem. Soc.* 10(3), 461–470 (2013)

[14] Sharma, C. D., Nagarkar, S. G., Eshwar, M. C.: Rapid spectrophotometric determination of scandium (iii) with (pyndylazo)- and (thiazolylazo) naphthols. *Bull. Chem. Soc. Jpn.* 59(5), 1662–1664 (1986)

[15] Shimizu, T.: Spectrophotometric determination of scandium with bromopyrogallol red. *Talanta* 14(4), 473–479 (1967)

[16] Shimizu, T., Ogami, K.: Analytical application and spectrophotometric study of the scandium indoferron complex. *Talanta* 16(12), 1527–1533 (1969)

[17] Swamidoss, C. M., Khopkar, S. M., Malkhede, D. D.: Liquid extraction and spectrophotometric determination of scandium (iii) with hexacetato calix arene. *Extr. Res. At. Miner.* 19, 211–218 (2009)

[18] Sydorova, L., Chernyavska, A., Chmylenko, T., Chmylenko, F.: Polyelectrolytic effect on spectrophotometric determination and interaction of scandium (iii) with eriochromcyanine. *East.-Eur. J. Enterp. Technol.* 4(6), 53–58 (2017)

4 Methods for spectrophotometric determination of zirconium (Zr)

4.1 Introduction

Zirconium, chemical element with atomic number 40, with atomic weight 91.22 is a metal of Group (IVB) of the periodic table. Before 1940, it was not much known, after which it became an important engineering material. The importance is due to its highly transparent property to neutrons, which makes it very suitable for nuclear reactors. In 1789, German chemist Martin Heinrich Klaproth identified the element Zr as in the orthosilicate Zr ($ZrSiO_4$). Later, the metal was isolated in 1824 in impure form by Swedish chemist Jons Jacob Berzelius. The metal is hard and brittle even when pure. The white, soft, malleable and ductile metal of higher purity was first produced in quantity in 1925. It was obtained by Dutch chemists Anton E. Van Arkel and J. H. deBoer by thermal decomposition of ZrI_4.

Zirconium is abundant in the Earth's crust, although not in concentrated deposits. The common mineral is zircon which is found in alluvial deposits, oceanic beaches, old lake beds and stream beds. Another important ore of zirconium is Baddeleyite which is pure zirconium dioxide ZrO_2. These zirconium minerals have some hafnium content, from a few tenths of a percent to several percent. In practice zirconium containing 1 % of hafnium is acceptable as pure zirconium.

The most important application of zirconium is in the nuclear reactors. The metal is used to clad fuel rods for alloying with uranium and also for reactor core structures since it has unique properties like good strength at elevated temperatures, corrosion resistance from rapidly containing coolants and no formation of highly radioactive isotopes, among others. It also withstands mechanical wear and tear from neutron bombardment. It absorbs oxygen, nitrogen and hydrogen in surprising large amounts. Its strong affinity for oxygen and other gases makes it extremely useful for removing residual gases in electron tubes. It has a number of other uses and applications that make it valuable in industry. In view of this, its spectrophotometric determination using various reagents comprises a significant part of analytical chemistry. In the subsequent sections, various methods published in last few years have published written to make analysts aware of its use in analytical spectrophotometry.

4.1.1 2-(2-thiazolyazo)-5-dimethyl-aminophenol (TAM) with zephiramine

Turumi C. [33] developed a spectrophotometric method for small amounts of zirconium using TAM as an analytical reagent. It described how a reddish violet complex of zirconium, which is stable for at least 70 minutes, is formed in pH range 3.7–4.5 at 595 nm. The zirconium complex with TAM and zephiramine in methanol is used for the deter-

https://doi.org/10.1515/9783111626833-004

mination. The Beer's linear range is up to 0.4 μg/mL of zirconium. The reported values of molar extinction coefficient and Sandell's sensitivity are 1.05×10^5 L/mol/cm and 8.69×10^4 μg/cm^2 per 0.001 absorbance unit using a masking agent. Interference from ions like copper hafnium, tungsten, tantalum, fluoride, oxalate, tartarate and citrate can be removed.

4.1.2 Sodium alizarin sulfate in perchloric acid

Mannug, D. L. et al. [37] developed a differential spectrophotometric method for the determination of zirconium using sodium alizarin sulfate in perchloric acid as a complexing reagent. An intensely red-colored complex was formed which absorbs at 530 nm. Beer's law range is 1.0 mg to 1.6 mg of zirconium. The coefficient of variation is reported to be 1 % comparable to gravimetric method. Fluoride, sulfate and phosphate are primary interfering ions as reported.

4.1.3 Phosphate solution

Kiefer, E. W. et al. [16] developed an indirect spectrophotometric method for the determination of zirconium. It was found by the authors that zirconium could be precipitated by standard phosphate solution and that the concentration of supernatant phosphate after centrifugation decreased in proportion to the zirconium present. This is used to determine zirconium spectrophotometrically using phosphate.

4.1.4 p (p-dimethylaminophenylazo) benzene arsonic acid DPB

Eberle, A. R. et al. [12] proposed a method to determine zirconium in thorium–zirconium alloys using DPB as a reagent. It was described that in HCl solution zirconium was precipitated by p (p-dimethyl-aminophenylazo) benzene arsonic acid, DPB. The free DPB, which is yellow when released in basic but red in acidic solution. Based on this reaction, the absorbance is measured at 560 nm against water as a blank. It further mentioned that, although thorium is also precipitated, it was to a small extent that can be controlled. Thus, the method is applicable to the determination of zirconium–thorium alloys.

4.1.5 4-(2-pyridylazo)-resorcinol

Nagarkar, S. C. et al. [20] developed a spectrophotometric method for the determination of zirconium using 4-(2-pyridylazo)-resorcinol as a reagent. The orange–red complex resulted from the reaction between zirconium and the reagent at pH 2.5 used to determine zirconium. It was described that in the presence of ascorbic acid and a phthalate buffer, the absorbance occurred at 540 nm. The Beer's law concentration range cited was 0.5 µg to 5.0 µg/ml. The molar-absorptivity value obtained was 6.75×10^3 and sensitivity 0.013 µg/cm^2 for zirconium. The method was successful with a standard deviation of ± 1.30 %.

4.1.6 8-hydroxyquinoline

Van Santen, R. T. et al. [35] described a method for determining small amounts of zirconium using 8-hydroxyaquinoline in chloroform as a reagent. The 8-hydroquinolate in chloroform absorbs at 386 nm, and the reported value of molar absorptivity was 14,700. It was mentioned that the method does not have any interference due to anions like Cl^-, SO_4^{2-}, perchlorate, tartarate, oxalate and fluoride.

4.1.7 Bromopyrogallol red

Sakai Takayoshi et al. [28] proposed a method for the spectrophotometric determination of zirconium using bromopyrogallol Red as reagent. It was documented that zirconium (iv) reacts with the reagent to form a violet complex which absorbs at 670 nm in the pH range 5.0–5.4. A stable complex is formed by boiling the solution for about 25 minutes, which sustains the color for 3 hours. The Beer's law range was 4–35 µg zirconium in 25 ml. The reported value of molar absorptivity was 7.01×10^4 at 670 nm, and the Sandell's sensitivity reported 0.002 µg/cm^2. The interference of diverse ions was also studied.

4.1.8 2-(5-bromo-2-pyridylazo)-5-diethylaminophenol

Rathaih, G. V. et al. [27] proposed a spectrophotometric method that is rapid using 2-(5-bromo-2-pyridylazo)-5-diethylaminophenol as a reagent. It was mentioned that zirconium reacts with the named reagent in pH range 3.8–5.8 to form a red chelate soluble in methanol–water mixtures. A 1:3 complex obeys Beer's law in the concentration range of 0.02–0.44 µg/mL of zirconium. Absorbance at 585 nm gives a molar absorptivity of 1.54×10^5 L/mol/cm. Interference by EDTA, Ga, In, Hf and V were reported to be serious.

4.1.9 2-(2-benzothiazolyazo)-3-hydroxyphenol

Alaa, S. Amin [8] developed a novel approach to determine zirconium by solid-phase spectrophotometry using 2-(2-benzothiazolylazo)-3-hydroxy-phenol as a reagent. Trace amount of zirconium (iv) in water, soil, plant or ore samples can be absorbed in a dextran-type, lipophilic gel as a complex with the reagent. The resin phase absorbance at 569 and 800 was directly used to determine Zr in the range 0.25–3.7 µg/L with a relative deviation of 1.06 %. The reported molar absorptivity was 1.08×10^5, 6.38×10^6 and

1.97×10^7 L/mol/cm at 100, 500 and 1000 mL, respectively. No significant interference was observed.

4.1.10 Mordant blue 2RL (MBR)

Mostafa M. Hamed et al. [19] developed a novel quick, simple yet cost- effective spectrophotometric method for the estimation of micro amounts of zirconium. The method is based on the reaction of Zr (iv) with mordant blue 2 RL (MBR) as a chelating agent. It was described that under optimum condition, MBR reacts with Zr (iv) in 2.0 M HCl to form a pink complex. The absorption maxima λ_{max} of the complex is at 583 nm and the Beer's law range is 0.0–12.0 µg/mL. Reported molar absorptivity was 4.93×10^3 L/mol/cm and Sandell's sensitivity 0.0185 µg/cm^2. The method is rapid, simple and does not a require skilled analyst.

4.1.11 Sulfobenzol azopyrocatechol with surfactants

I. A. Ahmedov [6] studied multi-ligand complex formation of zirconium with sulfobenzol azopyrocatechol (sulfobenzene azopirocatexin) and surfactants for the determination of zirconium. The λ_{max} was at 530 nm, and the complex was formed in an acidic environment at pH 3, when maximum zirconium ions form a complex. The Beer's law range was 0.24–6.5 µg/mL, and limit of detection was 3 µg/mL. It was mentioned that the factors that affect determination are volume of extraction, pH and the amount of ligand.

4.1.12 Diacetylmonoxime salicyl hydrazone (DMSH)

Anita Varghese et al. [36] developed a simple selective and sensitive spectrophotometric method for individual and simultaneous determination of trace amounts of vanadium (V) and zirconium (IV) in acetic acid medium with the reagent DMSH, without prior separation.

It was reported that zirconium complex absorbs at 380 nm with a molar absorptivity of 1.82×10^4 and a Sandell's sensitively of 2.5 ng/cm^2. A 1:2 complex with zirconium was observed using first-order derivative spectrophotometry, and both V (V) and Zr (IV) can be simultaneously determined at zero crossing wavelengths of 443 and 405 nm, respectively. The proposed method has been successful for the determination of zirconium in standard alloy steel samples, minerals and soils.

4.1.13 4-chloro-N-(2,6-dimethylphenyl)-2-hydroxy-5-sulfomoylbenzamide

Ahmed S. Al-Kady [7] developed a simple, selective and sensitive spectrophotometric method for the determination of trace amounts of Zr (IV) in aqueous samples. The method is based on complexation of Zr (IV) with 4-chloro-N-(2,6-dimethylphenyl)-2-hydroxy-5-sulfamoylbenzamide (Xipamide). In the ratio of 1:1, the Zr (IV) reacts with said reagent in pH 8 range to give a complex with absorption maxima at 333 nm. The Beer's law range was 0.2–3.6 µg/mL. The method was reported to be accurate and can be used for zirconium determination in the presence of a large number of ions. It could be successfully used for samples of zirconium in water, biological samples and soil samples, with a relative standard deviation of 0.195 %.

4.1.14 Dibenzo-18-crown-6

Y. K. Agarwal et al. [4] proposed an extraction-, separation- and preconcentration-based method for the spectrophotometric determination of zirconium. Zirconium (IV) was extracted with dibenzo-18-crown-6 from 0.1 M HCl and separated from hafnium. The system obeys Beer's law in the range 0.15–3.60 ppm of zirconium (IV) AL 283 nm. The molar absorptivity was 3.0×10^4 L/mol/cm. The method could be successfully applied to Zr (IV) determination in standard samples as well as sea-water samples.

4.1.15 Arsenazo III

S. Kus et al. [17] published a third-order derivative spectrophotometric method for the determination of zirconium using arsenazo III in strong acidic medium (4M HCl_{04}) and a 2-propanol-water (25 % V/V) medium. The method involves determination of the sum of chromium and zirconium at 637.4 nm. Oxalic acid is used to mask Zr and the value of the derivative at 673.4 gives. The concentration without any separation, by use of this method, can determine 5 µg of zirconium without separation, as described.

4.1.16 o-Nitrobenzene-azopyrocatechol (o-NBAP)

Amadov, Imtiaz [5] proposed a spectrophotometric method for the determination of zirconium using dispersive liquid micro extraction based on solidification of floating organic droplets. The complex formation was done by solidifying or floating organic droplets of zirconium o-NBAP and nonionic surfactant (PO-10). In the first step, separation and solidification are done, after which solidified samples are measured by spectrophotometric method in pH 2.5 and 510 nm. The Beer's law range is 0.5–70 µg/L. The method has been extended to the determination of zirconium in water samples.

4.1.17 Arsenazo III

Pang, Ming-Li et al. [25] proposed a method for the spectrophotometric determination of zirconium with a high mass concentration of lanthanum. Based on the priority of azo arsenic III to coordinate with H^+ in the solution of high acidity, the spectrophotometric determination of zirconium in the presence of La^{3+} was done. Th ratio of La^{3+} to Zr^{4+} was higher than 1.5. It was explained that La^{3+} is effectively masked when the concentration of HCl is higher than 0.3 mol/L. When the HCl concentration reaches 2.0 mol/L, the chromogenic reaction between Zr^{4+} and azo arsenic III showed the highest sensitivity. It was mentioned further that the curve equation for zirconium was $A = 0.10114C - 0.00117$ and correlation coefficient 0.9997. The reported Beer's law range was 0.4–2.4 µg/L. The authors claim ed that this method to be highly accurate.

4.1.18 DBC–arsenazo

Zhang, Xiao et al. [38] proposed a method for the spectrophotometric determination of zirconium using DBC–arsenazo as a reagent. It was described that Zr (IV) forms a 1:3 blue complex in 2.40 mol/L HNO_3 with DBCASA. The complex absorbed at 560 nm in the Beer's law range 0–1.20 µg/mL. The reported apparent molar absorptivity was 4.56×10^4 L/mol/cm with the limit of detection of 0.2 µg/mL. The method was used for the determination of Zr (IV) in water samples. Furthermore, the RSD value for 13 determination was 1.71–1.96 %, with a recovery of 95.8–104.1 %.

4.1.19 2,2',3'-4-tetra hydroxyl-3-sulpho-5'-carboxyazobenzene

Uysal, Ulku Dilek [34] published a spectrophotometric method for the determination of Zr (IV) in alloys using tetrahydroxyazon SC as a reagent. It was mentioned that tetrahydroxyazon SC in 5.0 M acidic medium (~ 6.0 pH) yields a red complex with Zr (IV). The complex absorbed at 495 nm under optimum conditions in 2 M HCl. The reported molar absorptivity was 4.9×10^4 L/mol/cm for a 1:2 complex, the Beer's Law range 0.10–2.50 mg/mL. The method was claimed to be rapid, simple and highly sensitive when applied to many alloys without separating in aluminum, zinc, copper etc.

4.1.20 Xylenol orange

Pourreza et al. [26] developed a method for the trace determination of zirconium using flotation–spectrophotometric technique. It was mentioned that this method is based on complex formation of zirconium with xylenol orange (Xo), which is floated in the interface of aqueous phase and n-hexane. When the aqueous solution and n-hexane solution discarded, the adsorbed complex on the wall of separating funnel is dissolved in small volume of methanol and is used for the determination. This complex absorbed at 429 nm in the linear range 7–120 ng/mL of zirconium. The reported limit of detection was 5.8 ng/mL, with a relative standard deviation for seven measurements for 50 and 110 ng/mL were 4.4 and 3.0 %, respectively. The method was extended to zirconium determination in water samples.

4.1.21 Polystyrene-2-oxi (azo-1)-2'-oxi-3',5'-dinitrobenzene

Basargin et al. [9] introduced a novel technique for the spectrophotometric determination of Zr (IV) in rock. The method was based on preliminary individual extraction (concentration) of Zr by a polymer chelating sorbent, polystyrene-2-oxi (azo-1)-2'-oxi-3',5'-dinitrobenzene.

The quantitative extraction of zirconium took place under static condition within acidic range of 1M HCl – 2.5 pH. The authors claim that the method showed has no interference with 6×10^4-fold of Na^+, K^+, Ca^{2+}, Mg^{2+}, Ba^{2+} and Sr^{2+}: even 2×10^4-fold of Al^{3+}, Cu^{2+}, Pb^{2+}, Co^{2+}, Na^{2+} and Cd^{2+} or Zn^{2+} are tolerated. 5×10^3-fold of Sn^{2+}, Fe^{2+}, Be^{2+}, and Ce^{2+} also do not interfere. A 2×10^6-fold mass amount of Fe^{2+}, Sc^{2+} and Y^{3+} does not interfere with the determination. The RSD value was reported as 0.01–0.04 at concentration of $n \times 10^{-2}$–$n \times 10^{-4}$ % of Zr (IV). The method was successful for Zr (IV) determination.

4.1.22 5,7-dibromo-8-hydroxyquinoline in presence of thiocyanate

Jain et al. [14] proposed a spectrophotometric method for determination of zirconium using 5,7-dibromo-8-hydroxyquinoline in presence of thiocyanate. The ternary complex formed among Zr (IV), the reagent and thiocynate is extractable into chloroform. The measurements were done at 416 nm to determine Zr (IV). The method obeyed Beer's law in a concentration range of 0.2–9.0 µg/mL of Zr. The reported values of molar absorptivity and Sandell's sensitivity were 1.05×10^4 L/mol/cm and 0.0087 µg/cm^2. The method showed no interference of a large number of foreign ions. The method proved successful for variety of zirconium samples.

4.1.23 *Platanus orientalis* tree laves acidic extract

Mongi, Akbar, Borein et al. [18] developed a novel method for the spectrophotometric determination of zirconium using a new reagent which is green as well. It was reported that the acidic extract of *Platanus orientalis* tree leaves in 6 M HCl reacts with zirconium yielded a yellow product. This complex showed a maximum absorbance at 422 nm, when the zirconium can be determined. The Beer's law range was 0.4–8 µg/mL and molar absorptivity 0.59×10^{-1} mol/cm. The reported value of the detection limit was 0.086 µg/mL. The authors claimed that the method was simple, clean, low-cost, sensitive and selective. It was applied for zirconium determination in varied samples like tap water, waste water and well water. An error of < 2.5 % was reported.

4.1.24 Salicylfluorone

Shimizu, Kumiyao et al. [30] developed a novel spectrophotometric method for the determination of zirconium (IV) and fluoride ion. The method was based on the complex formation between salicylfluorone, zirconium (IV) and fluoride ion in the presence of cationic surfactant, in the Beer's linear range of 0.090–0.46 μg/cm^3. The molar absorptivity at 550 nm was 1.7×10^5 L/mol/cm with a standard deviation of 0.32 % ($n = 6$). Further, the fluoride ion was detected at 552 in the vicinity of 3.865 ng/cm^3 in a strong acidic medium and the molar absorptivity was 2.1×10^5 L/mol/cm with RSD 1.3 % ($n = 6$).

4.1.25 2-(2-furyl)-3-hydroxy-4-oxo-4H-1-benzopyran

Agnihotri, Mehta [1] developed an extractive spectrophotometric method for the micro determination of Zr (IV) using a reaction with the said reagent. A dark yellow complex was formed by Zr (IV) in 1:4 (metal–ligand) ratio that can be extracted into benzene in the pH range 6.7–7.3. The complex absorbed at 420 nm, and Zr is determined with a standard deviation of 0.25 %. The method can be applied to several samples like flu dust as well as water samples.

4.1.26 Review of spectrophotometric methods for determination of zirconium

A voluminous review on various reagents and methods for the spectrophotometric determination of zirconium [10] has been published by the author of this book in condensed form, including methods reported from 1991 to 2004. Interested reader may consult this.

4.1.27 2,4-dihydroxy benzaldehyde isonicotinoyl hydrazone

Sivaramaiah et al. [31] proposed a spectrophotometric method for the determination of Zr (IV) with the mentioned reagent. The method is both direct and derivative. It was described that the Zr (IV) forms a golden yellow complex with 2,4-dihydroxy benzaldehyde isonicotinoyl hydrzone (2,4-DHBINH) at pH 1.5. This absorbed at 410 nm in the linear range 0.4–4.0 μg/mL of Zr (IV). The reported molar absorptivity and Sandell's sensitivity values were 1.82×10^4 L/mol/cm and 0.005 mg/cm^2. The method was tested against many interferences, and a method for a first-derivative mode of determination was also proposed.

4.1.28 3-hydroxy-2-(2'-thienyl)-4-H-chromen-4-one

Nijhawan et al. [23] proposed an extractive spectrophotometric method for the determination of trace amounts of Zr (IV). The analytical reagent 3-hydroxy-2-(2'-thienyl)-4H-chromen-4-one in ethanol was used for complexing and determination. It was reported that Zr (IV) and the reagent in acidic medium formed a colored species absorbing at 425 nm. In the Beer's aw range 0.0–2.2 µg/mL, the Sandell's sensitivity of 1.38×10^{-3} µg/cm^2 was observed. The authors claimed it to be applicable for a variety of zirconium samples.

4.1.29 Calixarene hydroxamic acid

Agrawal, Yadvendra et al. [3] synthesized and applied a new calixarene hydroxamic acid for the spectrophotometric determination of Zr (IV). The compound 25,26,27,28-tetrahydroxy-5,11,17,23-tetrakis (N-p-chlorophenyl) calyx (4) arene hydroxamic acid (CPCHA) was used in the extraction of Zr (IV). The complex absorbed at 447 nm with a molar absorptivity of 9.4×10^3 L/mol/cm, in the Beer's law range 1.0–9.5 ppm. The method has been applied to standard samples.

4.1.30 Lawsone (2-hydroxy-1,4-naphthaquinone)

Sedaira [29] published a method for the spectrophotometric determination of Zr (V) using Lawsone (2-hydroxy-1, 4-haphthaquinone) as a reagent. They described that the reagent forms a complex with Zr (IV) in 40 % (V/V) ethanol at 20 °C in an ionic strength of 0.1m ($NaClO_4$). A pH range 1.5–4.0 was optimum, and Zr $(Lawsone)_2$ absorbed at pH 3.3, 450 nm. The molar absorptivity reported was 1.13×10^4 L/mol/cm. An addition of cyanide masks several interfering species. The method was applied to many synthetic samples.

4.1.31 Thorin and N-hydroxy-N-N'-diphenylbenzamide

Deb, Manaskanti et al. [11] proposed a selective extraction-based spectrophotometric method for the determination of zirconium using thorin an N-hydroxy-N-N'-diphenylbenzamide (HDPBA). The method employed selective extraction of Zr (IV) with HDPBA in the presence of thorin as an ion-associated floated complex in a benzene–ethanol (3:7) mixture at pH 2.2. It was mentioned that most of the common ions associated with zirconium do not interfere. The method was more than 6-fold as sensitive compared to other methods. The complex (Zr (IV)–thorin-HDPBA) showed a λ_{max} around 495 nm. It was mentioned further that the complete extraction of this complex was achieved in presence of thorin with 2.78×10^{-3} mol/L to 3.4×10^{-2} mol/L, HDPBA at an optimum range of pH 1.0–2.2. The Zr contents determined by the method have been validated with AAs as claimed.

4.1.32 5-chloro-8-hydroxy-7-iodoquinoline

Nijhawan, Monica et al. [22] proposed a spectrophotometric method for the determination of trace amounts of zirconium using 5-chloro-8-hydroxy-7-iodoquinoline (chloroiodo-oxine) as a reagent. It was reported that a colored complex is extracted into chloroform from a dilute HCl solution. The complex absorbed at 410–415 nm. The reported values of molar absorptivity and Sandell's sensitivity were 1.57×10^4 L/mol/cm and 5.3×10^{-3} µg/cm^2, respectively, in the linear range of 0–6 µg/mL of Zr for 2 µg/mL of zirconium. The RSD value was 1.3 %, and a variety of other ions like Co (II), N (II), Mn (II), As (III), Bi (III), Ir (III), Au (III), Ru (III), Pt (IV), Th (IV), U (VI) and Os (VIII) did not interfere in the Zr determination.

4.1.33 5,7-diiodo-8-hydroxy-quinoline

Nijhawan, Monica et al. [21] proposed a method for the spectrophotometric determination of zirconium using 5,7-diiodo-8-hydroxy-quinoline, as a reagent. The reagent complexes with Zr (IV) in 1:2 ratio in dilute HCl produced yellow complex. The complex was extracted into chloroform from 0.1–0.2 m HCl showing an absorption a 417.5–421 nm. The reported molar absorptivity was 1.87×10^4 L/mol/cm, and the Sandell's sensitivity was 4.8×10^{-3} µg Zr/cm^2 in the linear range 0–5 µg/L Zr. The RSD value was I/-006 with relative mean error ± 0.36 %. The method was claimed to be simple, selective and sensitive.

4.1.34 Chrome Azurol S (CAS)

Ganesh, S. et al. [13] proposed a spectrophotometric method for the determination of zirconium using Chrome Azurol S (CAS) as complexing agent. The method was based on the fact that zirconium forms a bluish violet stable complex with CAS, which at pH 4.2 absorbs with a λ_{max} at 598. The Beer's law range was 1–7 µg/mL. The reported values of molar absorptivity were 3.93×10^3 L/mol/cm and the Sandell's sensitivity was 2.54×10^{-4} µg/cm^2. The limit of detection was 2.42×10^{-1} µg/mL. The method proved highly sensitive, selective and rapid and has been applied to zirconium in environmental, industrial, water and nuclear reprocessing samples.

4.1.35 4-(2-pyridylazo) resorcinol

Kalyanaraman et al. [15] studied a spectrophotometric determination method for zirconium using 4-(2-pyridylazo) resorcinol as complexing agent. It was stated that zirconium and hafnium react with PAR at pH 8–9.2 and 8–9.8, respectively, to yield 1:2 water soluble complexes. The molar absorptivity for zirconium was 66,000.

4.1.36 Sulfonitrazo Eh

Orlova, E. S. et al. [24] developed an improved method for zirconium determination using sulfonitrzo Eh as a reagent. This was developed to analyze alloy steel, and it makes possible the determination of zirconium without separation from other elements. It was noted that the zirconium complex had the following characteristics of a λ_{max} of 550 nm, a molar absorptivity of 1.34×10^4 L/mol/cm and a 1:1 complex.

4.1.37 N-phenylanthranilic acid and rhodamine B

Subramanian, M. S. et al. [32] proposed a spectrophotometric method for determining zirconium in steel samples. The method estimates 2–20 µg of zirconium using N-phenylanthranilic acid and rhodamine B to make an ion-association complex that is extractable in benzene. The extract absorbed at 560 nm and the molar absorptivity was 1.4×10^4 L/mol/cm.

4.1.38 N-p-chlorophenylbenzohydroxamic acid and morin

Agrawal, Y. K. et al. [2] proposed a spectrophotometric method for the determination of zirconium using N-p-chlorophenyl-benzohydroxamic acid (N-p-Cl-BHA) and morin. The method was based on the extraction of zirconium with N-p-Cl-BHA into isoamyl alcohol from 0.2–0.5 N H_2SO_4 and further complexing with morin. The greenish yellow complex absorbed at 420 nm and showed a molar absorptivity of 3.5×10^4 L/mol/cm. It was noted that the intensity as can be further enhanced by adding ammonium carbonate, which yielded a molar absorptivity of value 1.3×10^5 L/mol/cm at 410 nm.

A summary of analytical reagents used for the spectrophotometric determination of zirconium is described in Table 4.1.

Table 4.1: List of analytical reagents used for the spectrophotometric determination of zirconium.

S. No.	Name of the reagent	λ_{max} (nm pH/solvent)	Molar absorptivity (L/mol/cm)	Sandell's sensitivity (µg/cm^2)	Detection limit
1.	2-(2-thiazolylazo) 5-dimethylaminol-phenol	595 3.7–4.5	1.05×10^5	8.69×10^4	–
2.	Sodium alizarin sulfate in perchloric acid	530 Perchloric acid	–	–	–
3.	Phosphate solution	–	–	–	–
4.	p-(p-dimethyl-aminophenylazo) benzene arsonic acid (DPB)	560 HCl	–	–	–
5.	4-(2-pyridylazo) resorcinol	540 2.5	6.75×10^3	0.013	–
6.	8-hydroxy-quinoline	386 Chloroform	14700	–	–
7.	Bromopyrogallol Red	670 5.0–5.4	7.0×10^4	0.002	–
8.	2-(5-bromo-2-pyridylazo)-5-diehylamino phenol	585 3.8–5.8 Methanol	1.54×10^5	–	–
9.	2-(2-benzothiazolyl-azo)-3-hydroxy phenol	569 800	1.08×10^5 (100 mL) 6.38×10^6 (500 mL) 1.97×7 (1000 mL)	–	–
10.	Mordant Blue	583 2.0 M HCL	4.93×10^3	0.0185	–
11.	Sulfobenzol azopyrocatechol with surfactant	530	–	–	3 µg/mL
12.	Diacetylmonoxime salicyl hydrazone (DMSH)	380	1.82×10^4	2.5 ng/cm^2	–
13.	4-chloro-N-(2,6-dimethylphenyl)-2-hydroxy-5-sulfamoyl-benzamide (Xipamide)	333	–	–	–

Table 4.1 (continued)

S. No.	Name of the reagent	λ_{max} (nm pH/solvent)	Molar absorptivity (L/mol/cm)	Sandell's sensitivity (µg/cm²)	Detection limit
14.	Dibenzo-18-crown-6	283 0.1M HCl	3.0×10^4	–	–
15.	Arsenazo III	637.4 4M$HClO_4$	–	–	–
16.	O-nitrobenzene-azopyrocatechol(O-NBAP) & OP–10	510	–	–	–
17.	Arsenazo III	–	–	–	–
18.	DBC-arsenazo	560	4.56×10^4	0.22	–
19.	2,2'-4-tetrahydroxy-3'-sulpho-5'-carboxybenzene	495 6.0	4.9×10^4	–	–
20.	Xylenol orange	429	–	–	–
21.	Polystyrene-2-oxi (920-1)-2'-oxi-3',5'-dinitrobenzene	1M HCl	–	–	–
22.	5-7-dibromo-8-hydroxyquinoline & thiocyanate	416	1.05×10^4	0.0087	–
23.	Platanus orientalis tree leaves	422	0.59×10^{-1}	–	0.086 µg/mL
24.	Salicylfluorone	550 Strong acidic medium	1.7×10^5 2.1×10^5	–	–
25.	2-(2-furyl)-3-hydroxy-4-oxo-4H-1-benzopyran	420 6.7–7.3	–	–	–
26.	Review of number of reagents	–	–	–	–
27.	2,4-dihydroxy benzaldehyde isonicotinoyl hydrazone	410 1.5	1.82×10^4	0.005	–
28.	3-hydroxy-2-(2'-thienyl)-4-H-chromen-4-one	425	–	1.38×10^{-3}	–
29.	Calixarene hydroxamic acid	447	9.4×10^3	–	–
30.	Lawsone	450 3.3	1.13×10^4	–	–
31.	Thorin and N-hydroxy-N-N'-diphenylbenzenemide	495 2.2	–	–	–

Table 4.1 (continued)

S. No.	Name of the reagent	λ_{max} (nm pH/solvent)	Molar absorptivity (L/mol/cm)	Sandell's sensitivity (μg/cm^2)	Detection limit
32.	5-chloro-8-hydroxy-7-Iodoquinoline	410–415 Chloroform	1.57×10^4	5.3×10^{-3}	–
33.	5,7-diiodo-8-hydroxyquinoline	417.5–421 0.1–0.2 M HCl	1.87×10^4	4.8×10^{-3}	
34.	Chrome Azurol S	598 4.2	3.93×10^3	2.54×10^{-4}	2.54×10^{-4} μg/cm^2
35.	4-(2-pyridylazo) resorcinol	8–9.2	66000	–	–
36.	Sulfonitriazo Eh	550 –	1.34×10^4	–	–
37.	N-phenylanthranlic acid & Rhodomine B	560	1.4×10^4	–	–
38.	N-p-chlorophenyl hydroxamic acid & morin	420 0.2–0.5 N H_2SO_4	3.5×10^4 1.3×10^5	–	–

Bibliography

[1] Agnihotri, M.: A highly selective extractive spectrophotometric determination of zirconium (IV) using 2-(2-furyl)-3-hydroxy-4-oxo-4H-1-benzopyran. *J. Indian Chem. Soc.* 83(8), 846–848 (2006)

[2] Agrawal, Y. K., Patel, S. A.: Extraction and spectrophotometric determination of zirconium with N-p-chlorophenyl-benzohydroxamic acid and morin. *Bull. Soc. Chim. Belg.* 90(10), 1039–1044 (1981)

[3] Agrawal, Y. K., Sanyal, M., Shrivastav, P., Menon, S. K.: Sequential liquid-liquid extraction graphite furnace atomic absorption spectrophotometric (GFAAS) and spectrophotometric determination of zirconium (IV) with calixarene hydroxamic acid. *Talanta* 46(5), 1041–1049 (1998)

[4] Agrawal, Y. K., Sudhakar, S.: Extraction separation and preconcentration of zirconium. *Sep. Purif. Technol.* 27(2), 111–119 (2002)

[5] Ahmadov, I.: Spectrophotometric determination of zirconium by dispersive liquid-liquid micro extraction based on solid and cation of floating organic droplets. *Asian J. Chem.* 32(12), 3191–3196 (2020)

[6] Ahmadov, I. A.: Spectrophotometric determination of zirconium sulfobenzol azopyrocatechol and different ligand complexes with surfactants. *New Mater. Compd. Appl.* 5(3), 171–181 (2021)

[7] Al-Kady, A. S.: Selective and sensitive spectrophotometric determination of trace amounts of zirconium in environmental and biological samples using 4-chloro-N-(2,6-dimethylphenyl)-2-hydroxy-5-sulfamoylbenzamide. *Spectrochim. Acta, Part A, Mol. Biomol. Spectrosc.* 97, 284–289 (2012)

[8] Amin, S. A.: Novel approach for determination of zirconium by solid-phase spectrophotometry. *J. Taibah Univ. Sci.* 9(2), 227–236 (2015)

[9] Basargin, O., Smakov, R., Yu, G.: Concentration and spectrophotometric determination of zirconium (IV) during the analysis of rock using a chelating agent. *Inorg. Mater.* 46(15), 1637–1639 (2010)

[10] Dalawat, D. S., Chauhan, G.: Review of spectrophotometric methods for determination of zirconium. *Rev. Anal. Chem.* 24(2), 75–102 (2005)

[11] Deb, M. K., Nashine, N., Mishra, R. K.: Spectrophotometric determination of zirconium with thorin and N-hydroxy-N,N'-diphenylbenzamidine. *J. Indian Chem. Soc.* 74, 167–168 (1997)

[12] Eberle, A. R., Pinto, L., Lerner, M. W.: Spectrophotometric determination of zirconium in thorium–zirconium alloys. *Anal. Chem.* 34(9), 1176–1177 (1962)

[13] Ganesh, S., Velavendam, P., Pandey, N. K., Kamachi Mudli, U., Natarajan, R.: Spectrophotometric determination of zirconium using Chrome Azurol S (CAS) as complexing agent in aqueous streams of nuclear fuel reprocessing. *Int. J. Adv. Chem.* 1(1), 53–56 (2015)

[14] Jain, P. K.: Spectrophotometric determination of zirconium with 5,7-dibromo-8-hydroxyquinoline in presence of thiocyanate. *J. Anal. Chem.* 65(8), 820–824 (2010)

[15] Kalyanaraman, S., Fukasawa, T.: Spectrophotometric determination of zirconium and hafnium with 4-(2-pyrdylazo) resorcinol. *Anal. Chem.* 55(14), 2239–2241 (1983)

[16] Klefer, E. W., Boltz, D. F.: Spectrophotometric determination of zirconium. *Anal. Chem.* 24(3), 542–544 (1952)

[17] Kus, S., Obarski, N., Marczenko, Z.: Determination of thorium, uranium and zirconium with arsenazo III using third-order derivative spectrophotometry. *Anal. Sci.* 8, 213–218 (1992)

[18] Mongi, A. B., Zolfonoun, E., Ahmadi, S. J.: Application of acidic extract of *Platanus Orientalis* tree leaves as a green reagent for selective determination of zirconium. *Green Chem. Lett. Rev.* 1(2), 107–112 (2008)

[19] Mostafa, M. H., Refaat, F. A.: Simple, rapid and cost effective method for the determination of zirconium in different samples. *Microchem. J.* 149, 104032 (2019)

[20] Nagarkar, S. G., Eshwar, M. C.: Spectrophotometric determination of zirconium (IV) with 4-(2-pyridylazo)-resorcinol. *Mikrochim. Acta* 62, 797–800 (1974)

[21] Nijhawan, M., Chauhan, R. S., Kakkar, L. R.: Spectrophotometric determination of zirconium with 5,7-diiodo-8-hydroxyquinoline. *Bull. Chem. Soc. Jpn.* 68(10), 2885–2890 (1995)

[22] Nijhawan, M., Chauhan, R. S., Kakkar, L. R.: Spectrophotometric determination of zirconium with 5-chloro-8-hydroxy-7-iodoquinoline. *Chem. Anal.* 41(6), 991–997 (1996)

[23] Nijhawan, M., Kakkar, L. R.: Use of 3-hydroxy-2-(2'-thienyl)-4H-chromen-4-one as an analytical reagent for the spectrophotometric determination of zirconium. *Chem. Anal.* 44(4), 711–718 (1999)

[24] Orlova, E. S., Somova, T. N., Tikhonova, L. I.: Spectrophotometric determination of zirconium with sulfonitrazo Eh. *Zavod. Lab.* 49(6), 574–575 (1983)

[25] Pang, M.-L., Yang, X., Chen, Z., Li, Q., Zhang, H., Wu, C., Guo, L.: Spectrophotometric determination of zirconium with high mass concentration of lanthanum. *Mod. Chem. Ind.* 36(9), 201–203 and 205 (2016)

[26] Pourreza, P. S.: Determination of trace amounts of zirconium by flotation-spectrophotometric method. *J. Korean Chem. Soc.* 54(3), 283–286 (2010)

[27] Rathaih, G. V., Eshwar, M. C.: Rapid spectrophotometric determination of zirconium with 2-(5-bromopyndylazo)-5-diethylaminophenol. *Talanta* 35(6), 502–504 (1988)

[28] Sakai, T., Funaki, Y.: The spectrophotometric determination of zirconium (IV) with bromopyrogallol red. *Bull. Chem. Soc. Jpn.* 42, 2272–2275 (1969)

[29] Sedaira, H.: Spectrophotometric investigation of the complexation equilibria of zirconium (IV) with lawsone and determination of zirconium. *Monatsh. Chem.* 128(2), 147–156 (1997)

[30] Shimizu, K., Morita, Y., Tanaka, K., Kamino, S., Fujimoto, T., Yamaguchi, T.: Spectrophotometric determination of zirconium (IV) and fluoride ion using color reaction among salicylflurone, zirconium (IV) and fluoride ion. *Jpn. Anal.* 56(9), 773–779 (2007)

[31] Sivaramaiah, S., Raveendra Reddy, P., Krishna Reddy, V., Sreenivasulu Reddy, T.: Direct and derivative spectrophotometric determination of zirconium (IV) with 2,4-dihydroxybenzaldehyde isonicotinoylhydrazone. *Indian J. Chem, Sect. A, Inorg. Phys. Theor. Anal.* 42(1), 109–111 (2003)

[32] Subramanian, M. S., Rama Krishna, T. V.: Spectrophotometric determination of zirconium in steel with N-phenylanthranilic acid and Rhodamine B. *Indian J. Chem, Sect. A, Inorg. Phys. Theor. Anal.* 21(11), 1077–1078 (1982)

[33] Tsurumi, C.: Spectrophotometric determination of small amount of zirconium by 2-(2-thiazolylazo)-5-dimethylaminophenol (TAM) with zephiramine. *Jpn. Anal.* 26(4), 260–265 (1977)

[34] Uysal, U. D., Huseyinli, A. A., Gueray, T.: Rapid direct spectrophotometric determination of zirconium (IV) in alloys with 2,2',3-4-tetrahydroxy-3'-sulpho-5'-carboxyazobenzene reagent. *J. Sci. Ind. Res.* 70(1), 45–50 (2011)

[35] Van Santen, R. T., Schlewitz, J. H., Toy, C. H.: The spectrophotometric determination of zirconium with 8-hydroxyquinoline. *Anal. Chim. Acta* 33, 593–596 (1965)

[36] Varghese, A., George, L.: Simultaneous first order derivative spectrophotometric determination of vanadium and zirconium in alloy steels and minerals. *Spectrochim. Acta, Part A, Mol. Biomol. Spectrosc.* 95, 46–52 (2012)

[37] White, J. C., Manning, D. L.: Differentiated spectrophotometric determination of zirconium. *Anal. Chem.* 27(9), 1389–1392 (1955)

[38] Zhang, X.-X., Zhao, Y., Wang, S., Zhai, Q.: Spectrophotometric determination of zirconium in water samples. *Asian J. Chem.* 25(1), 587–588 (2013)

5 Methods for spectrophotometric determination of titanium (Ti)

5.1 Introduction

Titanium, atomic number 22, is member of Group 4 (IVB) of the periodic table. It is a lightweight, low-corrosion and high-strength metal. Titanium, due to these properties, is used in alloys to make parts of high-speed aircrafts. Titanium constitutes 0.44 percent of Earth's crust and is found in rocks, clay, sand and many other soils. Plants and animals also contain titanium. Titanium is formed in mainly two minerals: ilmenite and rutile. Mathew A. Hunter first isolated it in pure form. Pure titanium is about half as dense as iron and less than twice as dense as aluminum, has a very low electrical conductivity, is thermally conductive and is weakly paramagnetic. Its application as an alloying agent is well-established. It is resistant to corrosion and thus tolerates many types of environmental conditions since it forms a passive oxide film. At elevated temperatures, it is reactive with oxygen in the air. In the liquid state, it is very reactive and reduces all known refractories. Titanium is an extremely versatile metal and thus has diverse applications. The high strength-to-weight ratio, corrosion resistance and biocompatibility makes it an important metal. which has found applications in aerospace, marine, automotive, medical, industrial even sporting goods. Consequently, both its application and the determination or analysis of titanium comprise even a wider part of the mentioned areas. In the preceding section, the spectrophotometric methods using various analytical reagents for titanium have been described. Although care has been taken to be chronological, the listed articles and their publication years may not be in exactly sequential. Table 5.1 concludes this chapter with a complete list of analytical various reagents and related details for titanium determination methods.

5.1.1 Ascorbic acid

Hines E. et al. [13] developed a spectrophotometric method for the determination of titanium using ascorbic acid as a reagent. It is noted that with a large excess of ascorbic acid in the pH range 3.5–6, a stable complex is formed having λ_{max} at 360 nm. The linear range for Beer's law is 0.1–25 ppm. The procedure is suitable for the determination of titanium as verified by standard samples. It was further mentioned that ions such as fluoride, phosphate and silicate mainly interfere with the determination.

5.1.2 Methylene Blue and ascorbic acid

Zavvar, H. et al. [28] developed a highly sensitive and selective catalytic method for the spectrophotometric trace determination of titanium (IV). The method is based on the catalytic effect of Ti (IV) on the methylene blue-ascorbic acid-redox system. This reac-

https://doi.org/10.1515/9783111626833-005

tion was followed spectrophotometrically by measuring the change in the absorbance of methylene blue at 665 nm. The Beer's law range was 3–25 ng/mL of Ti (IV). The RSDs for 10 and 20 ng/mL of Ti (IV) were 2.64 % and 1.55 % for 8 readings, and the detection limit was 0.6 ng/mL. Further, the method has been applied to the Ti (IV) determination in ore samples and tap water.

5.1.3 Hydrogen peroxide and xylenol orange

Mokoto, Otomo [29] proposed a spectrophotometric method for the determination of titanium using hydrogen peroxide and xylenol orange at pH below 2. When titanium is treated with a solution of xylenol orange at pH < 2, almost no reaction yielding color is observed. The addition of hydrogen peroxide as a third reactant gives red-colored complex which absorbs at 535 nm. The color is stable, yet it is affected by the amount of hydrogen peroxide. When 0.2–5 ml of a 0.3 % solution of H_2O_2 and 3.5–6 ml of a 1×10^{-3} M solution of xylenol orange were used for < 70 µg titanium and a total volume of 25 mL, the color obtained was stable for 30 minutes. When the measurements were made, 7 to 70 µg titanium could be measured in the linear range. A 1:1:1 complex is reported.

5.1.4 Diantipyrinylmethane

Wood, D. F. et al. [40] used the reaction between titanium and diantipyrinylmethane to form a yellow-colored complex for the spectrophotometric determination of titanium. In the range 20–250 ppm, titanium can be determined by measuring at 390 nm in a mineral acid solution. This method is claimed to be more sensitive than the peroxide method, and it is less susceptible to interference from other metal ions. The method can be applied to ZR 30 alloys without separation of molybdenum. Further, in the presence of appreciable amounts of molybdenum and niobium, the measurements were made at 430 nm.

5.1.5 Salicylic acid in a H_2SO_4 medium

Hultqust, A. E. [16] studied the spectrophotometric determination of titanium using salicylic acid in a H_2SO_4 medium. It was reported that a stable 1:1 complex was formed which absorbed at 410 nm. The molar absorptivity reported was 2.5×10^3 L/mol/cm. This method had no interference from several ions.

5.1.6 Nitric acid and hydrogen peroxide

Xian Feng, Du et al. [8] proposed a simple and rapid spectrophotometric method for titanium determination on etched aluminum forts. It has been described that after optimizing variables like wavelength, nitric-acid concentration, and H_2O_2 concentration in the concentration range 0.01–3.0 m mol/L, titanium can be determined with accuracy. The absorbance was 410 nm and RSD ranged from 0.67 % to 1.09 % for $n = 6$. The results obtained are in good agreement with those obtained by ICP-AES.

5.1.7 Thiocyanate

Viasta Vojkovic et al. [39] developed a spectrophotometric method for the determination of Ti (IV) using thiocyanate as a reagent. The method was based on the extraction of Ti (IV) as an ion-associate complex formed between titanium (IV)–thiocynate and cetyltrimethyl ammonium (CTA), cetylpyridinium (CP) or tetradecyl-dimethylbenzylammonium (TDMA) cation in chloroform or dichloromethane. It was reported that the systems studied were most sensitive and absorbed at 420 nm for measurements. The molar absorptivity in the presence of chloride ions were $6–7 \times 10^4$ L/mol/cm and 20–35 ng Ti (IV)/mL. The method could tolerate a number of ions without interference. The validation of results was done to determine titanium in bauxite and aluminum samples.

5.1.8 N'-(2-hydroxybenzylidene)-3-oxobutanehydrazide

Srilalitha, V. et al. [32] proposed a new, rapid and simple method for the precise determination of Ti (IV) using N'-(2-hydroxygenzylidene)-3-oxobutanehydrazide as a chromogen. It was mentioned that the complex at pH 2.0, absorbs at 500 nm in the linear range of 1.757–17.54 µg/mL. The reported values of molar absorptivity and Sandell's sensitivity were 1.68×10^4 L/mol/cm and 0.00285 µg/cm^2, respectively. The reproducibility of the method was excellent, and the detection limit was 0.385 µg/mL with an RSD of 1.595. The method was applied to titanium (IV) determination in synthetic mixtures and alloys.

5.1.9 Tetraphenylarsonium or tetraphenylphosphonium thiocyanate

Tamhina, B. et al. [34] proposed a spectrophotometric method for the determination of titanium (IV) with thiocynate by extracting the complex as a tetraphenylarsonium and tetraphenylphosphonium ion-associate complex. The chloroform extract absorbed at 420 nm in the Beer's law range 0.1–1 µg Ti/mL. This complex remained stable for 5 hours with a molar absorptivity of 7.5×10^4 L/mol/cm. Further. the complexes extracted were also studied in the solid-state.

5.1.10 Alizarin and fluoride

Lopez, R. Nunez et al. [22] developed a simple spectrophotometric method for determining titanium (IV) using a mixed-ligand Ti-fluoride-alizarin complex. The complex was formed at pH 9.5–10.3 with alizarin in the presence of fluoride that can be extracted into methyl isobutyl ketone and is red–violet. The complex absorbed at 513 nm, and the Beer's law linear range ranged up to 22 µg of Ti (IV) in 30 mL. The molar absorptivity was 7.0×10^4 L/mol/cm. The method has been used to determine titanium in an oxide mixture as well as aluminum alloys.

5.1.11 p-Hydroxybenzoylhydrazone (RAPHBH) in presence of Tritan X-100

Varghese, Anita et al. [38] developed a simultaneous determination method for Ti (IV) and Mo (VI) using RAPHBH as a reagent. The complex is used for the spectrophotometric determination in the presence of Triton X-100 as a surfactant. For Ti (IV), the Beer's law range was 0.13–1.2 µg/mL and λ_{max} 455 nm. The colored complex at pH 3 and 455 nm had a molar absorptivity value of 3.1×10^4 L/mol/cm, whereas the Sandell's sensitivity was 1.6 ng/cm^2 for Ti (IV). In addition to this, the authors noted that the first-derivative spectra of these complexes permitted simultaneous determination of Ti (IV) and Mo (VI) at zero-crossing wavelengths of 500 and 455 nm, respectively. The method proposed has successfully determined titanium and molybdenum in standard alloy steels, minerals and soil samples.

5.1.12 2,4-dihydroxybenzaldehyde isonicotinoylhydrazone

Babaiah, O. et al. [2] proposed a simple, sensitive spectrophotometric method for the determination of titanium in aqueous media. The reagent 2,4-DHBINH in the pH range 1–7 formed a reddish-brown complex with titanium, which absorbs at 430 nm and 500 nm. However, the reagent shows appreciable absorbance at 430 nm and negligible absorbance at 500 nm at pH 1.5. The Beer's law range was 0.09–2.15 µg/mL, and the reported values of molar absorptivity and Sandell's sensitively were 1.35×10^4 L/mol/cm and 0.0049 µg/cm^2, respectively. The authors also proposed a first-order derivative spectrophotometric method for titanium. The method proved to be valid for titanium determination in several alloys and steel samples.

5.1.13 Chromotropic acid and oxalate

L. F. Capitan Vallvey et al. [5] proposed a micro determination method for titanium in seawater using ion-exchange spectrophotometry. Titanium reacted with chromotropic acid oxalate to form a ternary complex which can be absorbed on Dowex 1-X8 anionic resin. The resin phase was directly measured at 440 and 800 nm. The method had a detection limit of 0.6 µg/l for a 500 mL sample and 0.4 µg/l for a 1,000 mL sample. The RSD value was 4 %. This method has been applied for the determination of titanium in seawater samples on the Mediterranean coast.

5.1.14 2,3-dihydroxynaphthalene

Mondal, R. K. et al. [26] developed a spectrophotometric method based on a chelate of 2,3-dihydroxynaphthalene with titanium. At pH 4–9, it forms a strong chelate which can be extracted into ethylacetate along with iron, leaving behind V, Nb, Mo of silicate rocks. Then, titanium can be separated from iron by raising pH and re-extracting. The report cites the molar absorptivity as 3.2×10^4 L/mol/cm. The method has applications in extractive the spectrophotometric determination of titanium in silicate rocks soils and columbite–tantalite minerals.

5.1.15 2,3-dihydroxynaphthalene (H_2ND)

Tarafdar, P. K. et al. [35] developed a sensitive, yet highly selective, extractive and aqueous-based method for determination of titanium (IV). The chelating agent used is 2,3-dihydroxynaphthalene (H_2ND). The reagent formed a 1:3 complex at pH 4–9. The reported molar absorptivity and Sandell's sensitivity values were 3.2×10^4 L/mol/cm and 0.001 μg/cm^2 at 375 nm. The selectivity for the method was very high, particularly for rock samples, and the solvent extraction using ethyl acetate improved the detection limit. This method has successfully determined titanium in silicate rock samples in terms of sensitivity, selectivity, simplicity and economy as claimed.

5.1.16 Haematoxylin and cetyltrimethylammonium bromide

Leong, C. L. [21] studied and developed a spectrophotometric method for the determination of titanium (IV) using haematoxylin and cetyltrimethylammonium bromide as ligands. The ternary complex had a molar absorptivity of 3.5×10^4 L/mol/cm after 1/2 hours of color development. At level the of 1 mg, no interference from tungsten (VI), aluminum (III), silver (I), lead (II), copper (II), zinc (II) or cadmium (II) was observed. However, interference from molybdenum (VI), antimony (V), chromium (VI), niobium (V), mercury (II) and manganese (II) at 1-mg level and iron (III) and tin (IV) at 0.5-mg level can be eliminated by pre-extraction of titanium and a titanium–cupferron complex using 4-methylpentan-2-one.

5.1.17 N-phenyl laurohydroxamic acid and phenyl fluorine

H. Pasaratha Gunawardhan [11] proposed an extraction spectrophotometric method for the determination of Ti (IV) in a 9–10 M HCl to yield a complex with the mentioned reagents extractable in hexane or chloroform. It was reported that chloroform extracts the titanium complex. When extracted for second time in a dilute HCl (0.1–0.5 M) in the presence of phenyl fluorine and isoanyl alcohol, there formed a red-colored complex that absorbs at 540 nm. The molar absorptivity at 540 nm was 2.33×10^5 l/mol/cm, and at 560 nm 1.23×10^5 l/mol/cm up to 0.4 ppm, the linear range was used to make the determination. The method can be used to determine titanium presence at 10 parts/10^9.

5.1.18 N1-hydroxy-N1,N2 diphenyl-benzamidine and thiocyanate

Yirgalem, Y. et al. [41] developed a precise selective and reliable extraction- based spectrophotometric method for the determination of titanium [22]. The mentioned reagent reacts with titanium and thiocyanate to produce an orange–red complex extractable into toluene from 0.05–0.15 M HCl. This complex absorbed at 400 nm and had molar absorptivity of 20,000 l/m/cm, in the Beer's law range of concentration 0.20–3.0 µg/mL. The RSD was 0.53 %, and method has been applied to titanium containing ores, minerals and alloys.

5.1.19 Tannin and thioglycollic acid

Samaresh, B. [4] developed a sensitive spectrophotometric method for the determination of titanium by complex formation with tannin and thioglycolic acid at pH 4.0. The complex absorbed at 400 nm in the Beer law range of 0.2–5 ppm of titanium. The interference of Co, Ni, Cr, Mn, V, Mo and W is eliminated by the polar separation of titanium by controlled addition of cupferron in the presence of thioglycolic acid (TGA). Copper was separated by precipitation with TGA, yet niobium interfered in traces. It has been used for determination in steels, magnet alloys etc.

5.1.20 Diantipyrylmethane with ascorbic acid

Feng-Ying Zheng et al. [42] developed a novel method for preconcentration of Ti (IV) using nanosized ZrO_2 and the spectrophotometric determination of Ti (IV). It was explained that Ti (IV) was selectivity absorbed on 300-mg ZrO_2 from a 500-mL solution at pH 6.0, then eluted by adding 5 mL, 11.3 mol/L of HF. The element added was itself the chromogenic reagent DAPM, and the masking agent ascorbic acid also was used. This system absorbed at 390 nm, with a detection limit (3σ $n = 11$) of 0.1 µg/L. The method was successful for the determination of Ti (IV) in river water and seawater also.

5.1.21 Tiron

A. V. Moharir et al. [25] developed a procedure for the spectrophotometric determination of titanium using tiron (disodiuim 1,2-dihydroxybenzene-3,5-disulfonate) and modifying it by thioglycolic acid. This reduces the Fe (III)–tiron complex which otherwise interferes with the measurements. It was noted that the reduction by thioglycolic acid was instantaneous, and reoxidation does not occur. A pH 5.0 buffer is used to stabilize the system. No interference from large amounts of Mn, Zr, Al, P, F or citrate was observed. More than 0.2 ppm of Cu and Ni are not tolerated and P and Zr should not be present together.

5.1.22 2,4-dihydroxy acetophenone benzoylhydrazone

Jam, Attimukat et al. [17] proposed a simple, sensitive spectrophotometric method for the trace determination of titanium. 2,4-dihydroxyacetophenone benzoylhydrazone was used as a complexing agent in a neutral medium. It was described that the complex so formed was quantitatively extracted into chloroform and used for the measurements. The complex showed λ_{max} at 371 nm in the linear range 0.0–3.5 µg Ti/mL. The reported values of molar absorptivity and Sandell's sensitivity were 1.35×10^4 L/mol/cm and 0.0035 µg/cm^2. The detection limit was 0.0146 ppm with a standard deviation of 0.0012 and an RSD of 0.276 %, respectively. The method as claimed was highly selective, reproducible and has been used for a large number of samples with great accuracy.

5.1.23 Chromotropic acid disodium salt and hexaacetato calyx (6) arene

Swamidoss, Caroline, M. A. et al. [33] proposed a method for the extraction- based spectrophotometric determination of titanium (IV) using hexaacetato calyx (6) arene as an extractant. The determination basically uses a chromotropic acid disodium–Ti complex which is extracted at pH 3. The organic phase when 1×10^{-4} M of ligand in xylene at pH 3.0 with the ligand shows absorbance at 470 nm. The titanium could be determined in real samples.

5.1.24 5-bromo-2-hydroxyl-3-methoxybenzaldehyde-p-hydroxy-benzoic hydrazone

Devi Reddy, M. et al. [7] proposed a method for the spectrophotometric determination of Ti (IV) using a thick orange-colored soluble complex with this reagent. In aqueous dimethyl formamide, this complex has its λ_{max} at 390 nm. The determination was carried out at pH 4.0, and the system obeys Beer law at 0.241–2.85 µg/m. The molar absorptivity was 1.44×10^4 L/mol/cm, Sandell's sensitivity was 0.0033 µg/cm^2 and the standard deviation ($n = 10$) was 9.04×10^{-3}. The results agreed well with certified values.

5.1.25 Diantipyrylmethane

Ma, Guang et al. [23] proposed a spectrophotometric determination method for titanium in high-bearing Ti blast-furnace slag using diantipyrylmethane. It explained that due to low reactivity and difficult dissolution of the blast- furnace slag, the ball needs to have been milled below 100-mesh. After treatment. the melt was leached with sulfuric acid under heating, and ascorbic acid was used to mask Fe^{3+}. The final content in high-Ti bearing blast furnace slag was determined using diantipyrylmethane spectrophotometrically. This complex absorbed at 370 nm and had an apparent molar absorptivity of 1.06×10^4 L/mol/cm, in the concentration range of 0.05–20 µg/mL. The RSD value was 0.24–0.57 % ($n = 5$).

5.1.26 2,3-dihydroxynaphthalene

Thakur, R. et al. [37] proposed an extraction-based spectrophotometric method for the determination of Ti (IV) and molybdenum. A facile liquid–liquid extraction for titanium and molybdenum used 2,3-dihydroxynaphthalene as an extractant in n-butylacetate at pH 4–9. The n-butyl extract of Ti (IV) absorbed at 390 nm, and the molar absorptivity of Ti-system was 3.7×10^4 L/mol/cm. The method has been applied to samples of silicate rocks, ores, minerals etc.

5.1.27 o-Carboxyphenylflurone in cationic micellar media

Hoshino, Mitsuru et al. [14] proposed a spectrophotometric method for the determination of Ti (IV) using o-carboxyphenylflurone in the presence of hexadecyltrimethyl ammonium chloride (HTCA) in a strongly acidic medium. In the Beer's law linear range 24–34 ng/mL of titanium, the complex absorbed at 530 nm. The molar absorptivity was 2.24×10^5 dm^3/mol/cm, and RSD 0.64 % for $n = 8$. The addition of EDTA eliminated serious interference of iron. The complex formed was an orange-colored Ti $(OCPF)_4$, whose reaction conditions and other parameters have been investigated by the authors.

5.1.28 4-hydroxyderricin (Hy) or xanthoangelol (Xa)

Harada, Aki et al. [12] proposed a new, simple yet sensitive spectrophotometric method for the determination of Ti (IV). The method used complex formation between titanium and 4-hydroxyderricin (Hy) or xanthoangelol (Xa) in the presence of an anionic surfactant. It was mentioned that in the linear range 0.01–0.96 µg/mL of Ti (IV) in a strong acidic medium, the complex absorbed at 433 nm. The RSD was 0.53 % ($n = 6$). The method has been applied to titanium determination in the Japanese herb Ashitaba.

5.1.29 Thiocyanate and cetyl trimethylammonium bromide

Tarafdar, P. K. et al. [36] proposed a highly sensitive spectrophotometric method for the determination of titanium in silicate rocks using thiocyanate and a cetyltrimethylammonium bromide ternary complex. The 1:2:2 (M: SCN^- CTA) complex absorbed at 421 nm. A deep yellowish complex was formed which can be extracted from an aqueous solution having an HCl and SCN^- concentration in 1–5 mol/L and 1.5–2.5 mol/L, respectively. The reported molar absorptivity was $1.1–1.0\times10^5$ L/mol/cm, and the Sandell's sensitivity was 0.47 ng/cm^2 at 421 nm. The method was reported to be highly sensitive and selective and has been applied for estimating titanium in a number of silicate rock samples. The results obtained are accurate, reproducible and comparable to reference materials.

5.1.30 Diantipyrylmethane

Gao, Liu [10] proposed a method for the spectrophotometric determination of titanium in aluminum alloys by using diantipyrylmethane as a complexing agent. It was reported that the alloy sample was rapidly decomposed by HNO_3–HF, and excess HF was neutralized by H_3BO_3. The color reaction occurred in the medium of 1.2 m/L HCl–1.8 M/L H_2SO_4, and in this medium other coexisting ions did not interfere. In the linear range 0.40–10.0 µg/L of Ti (IV), at around 400 nm, Ti (IV) could be determined. A standard deviation of 2.5 %–5.0 % was reported by the authors.

5.1.31 o-Nitrophenylfluorone and H_2O_2

Zhou, Zhi-Rong et al. [43] proposed a spectrophotometric method for the determination of titanium (IV) based on the catalytic effect of Ti (VI) on the oxidation reaction of o-nitrophenylfluorone (o-NO_2-PF) and H_2O_2 in a 0.016 mol/L H_2SO_4 medium at 90 °C. This catalytic effect was monitored spectrophotometrically by a decrease in absorbance of o-NO_2-PF at a wavelength of 550 nm after heating for 8 minutes. The linear range for the determination was 0–80 µg/L of Ti (IV), and detection limit was 0.064 µg/L. with an RSD value reported for 0.2 µg/25 mL titanium as 1.86 % (n = 11). The results agreed well with certified values.

5.1.32 Chromotropic acid

Kika, Fotmi, S. et al. [19] developed a first sequential injection (SI) method for TI (IV) determination spectrophotometrically. The basis of the method is the complex formation between titanium (IV) and chromotropic acid in an acidic medium forming a water-soluble complex. The complex absorbed at 420 nm. The reported linear range was 0.2–1.0 mg/L of Ti (IV) with a detection limit of 0.7 mg/L (n = 10). It is a selective method satisfactorily applied to the determination of titanium in dental implants and natural Moroccan rock phosphate.

5.1.33 Water extract of slippery elm leaf

Monji, Akbar Boveiri et al. [27] proposed a novel spectrophotometric method for the determination of titanium (IV) using a water extract of slippery elm leaf. A 0.05 M solution of Ti (IV) reacted with this reagent to form a yellow product. The product absorbed at 415 nm, and Ti could be determined by using this wavelength. The linear range of determination was 0.2–6 µg/L, and molar absorptivity was 0.68×10^4 mol^{-1} cm^{-1}. The reported detection limit was 0.0131 µg/mL. The proposed method was claimed to be simple, lost-cost, selective and sensitive.

5.1.34 4-(2-thiazolylazo)-1,2,3-trihydroxybenzene

Ma, Wei-Xing et al. [24] developed a spectrophotometric method for titanium (IV) determination in steel using 4-(2-thiazolyazo)-1,2,3-trihydroxybenzene as a reagent. It was described that in $ClCH_2COOH$-$ClCH_2COON$ a buffer at pH 2.0–5.0 titanium reacts with TATB in the presence of CTMAB to form a stable red complex in 1:2 (Ti:TATB) ratio. This complex absorbed at 560 nm and had an apparent molar absorptivity of 4.07×10^4 L/mol/cm. The Beer's law range was 0–20 µg/25 ml of Ti (IV). The method has been applied successfully for the determination of Ti (IV) in steels, and the results obtained agree well with the certified value.

5.1.35 2-hydroxy-3-methoxyphenol-fluorone (HMPF)

Chang, Shi-Ke et al. [6] developed a new method for the spectrophotometric determination of titanium in alloy steel based on the color reaction of Ti (IV) with HMPF. In a H_2SO_4 medium and in the presence of OP, HMPP reacted with Ti (IV) to give a 1:3 red complex that absorbed at 536 nm. The apparent molar absorptivity of this complex was 1.9×10^5 L/mol/cm in the linear concentration range 0–3 µg/25 mL of Ti (IV). The method has been successfully applied to determine Ti (IV) in alloy steels without prior separation and with satisfactory results.

5.1.36 Bromopyrogallol red in an ionic/nonionic micellar surfactant-mixed micellar media

Huang, Xiong et al. [15] proposed a method for the spectrophotometric determination of Ti (IV) using bromopyrogallol red as a reagent. It was suggested that Ti (IV) inhibits the decolorization reaction of bromopyrogallol red in a mixed micellar medium that contained cetyltrimethylammonium bromide (CTAB) and nonylphenoxypolyethoxyethanol (OP). This effect gives rise to a kinetic spectrophotometric method for Ti (IV) determination. The method is quite sensitive, with a molar absorptivity of 4×10^6 mol/l/cm, and the limit of detection was 0.2 ng/cm^2.

5.1.37 4-nitrocatechol and tetrazolium salts tetrazolviolet and iodonitrotelrazolium chloride

Kostova et al. [20] proposed a method of spectrophotometric determination for Ti (IV) using the cited reagents. It was described that the extraction of the ion-associates of Ti (IV) with 4-nitrocatechol and the tetrazoviolet and iodonitrotetrazolium chloride formed a 1:2:2 complex. Beer's law was followed in the range 0.172–2.044 $\mu g/cm^2$ for ion-association complexes with INT and 0.20–200 $\mu g/cm^2$ with TVD. The complex associate absorbed at 430 nm, with a molar absorptivity of INT 2.38×10^4/mol/l/cm at 440 nm. The method has been applied to aluminum–titanium alloys.

5.1.38 Thiocyanate

Khan, Kamin et al. [18] proposed a rapid and sensitive method for the separation and determination of titanium and ions in thiocyanate system. They examined the extraction of Ti (II) by a high molecular-weight amine in the organic solvent ($CHCl_3$) and subsequent separation from Fe (III). This was shown that the Ti (IV)–SCN complex was quantitatively extracted into an organic phase containing tribenzylamine, while the blood red Fe (III)–SCN complex gets separated in the aqueous phase. The molar absorptivity of Ti (IV)–SCN complex was 1.9×10^5 L/mol/cm. This method has been successfully applied to titanium determination in Cherat volcanic ash.

5.1.39 3,4-dihydroxybenzoic acid

Ferreira, S. L. C. et al. [9] developed a fast spectrophotometric method for the determination of titanium in geological matrices. They described how Ti (IV) ions form a yellow complex with a sample solution containing the chromogenic agent 3,4-dihydroxybenzoic, an acetate buffer and ascorbic acid. The complex absorbed at 380 nm. The results reported show that in the pH range 4.0–5.0, the complex exhibited a molar absorptivity of 1.43×10^4 l/mol/cm with a detection limit of 2.3 ng/mL. This method makes possible analysis of 30 samples/hour, and common anions and cations do not interfere, even in large numbers. The interference of iron (III) was eliminated by ascorbic acid. The procedure has been successfully applied to titanium determination in different geological samples, with precision (RSD $< 1\%$).

5.1.40 5-chlorosalicylic acid

Sedaira, Hassan et al. [31] proposed a spectrophotometric method for titanium (IV)— using 5-chlorosalicylic acid as a reagent. The Ti (IV)–CSA complex formed in the pH range 25–30 and absorbed at 355 nm in the linear range 0.53–4.8 µg/mL. The molar absorptivity was 1.9×10^4 l/mol/cm and Sandell's sensitivity 1.25×10^{-3} μg/cm^2. The limit of detection of titanium was 12 ng/mL. Ascorbic acid was used to mask Fe (III).

5.1.41 3-(o-carboxyphenol)-1-o-tolyltriazene-N-oxide: triazene

Banerjee, N. M. et al. [3] used 3-(o-carboxyphenyl)-1-o-tolyltriazene-N-oxide as a chromogenic reagent for the spectrophotometric determination of titanium (IV). The reagent formed a 1:2 yellow-colored complex with Ti (IV) in pH range 1–3.5 showing a λ_{max} at 410 nm. The reported molar absorptivity was 5.74×10^3 l/mol/cm in the Beer's law range of 0.25–65 ppm. Sandell's sensitivity was 8.3 ng/cm^2.

5.1.42 2-pyridineiminobenzohy-droxamic acid (2-PIBHA)

Salmas, F. et al. [30] first used the reaction of Ti (IV) with 2-PIBHA in an aqueous medium and extraction in orange alcohol for the spectrophotometric determination of Ti (IV). It was mentioned that a 1:1 and 1:3 (Ti:Reagent) are formed in an aqueous solution and extracted in isoamyl alcohol which absorbs at 375 nm. In the linear concentration range of 0.1–1.4 μg/mL of Ti (IV), the apparent molar absorptivity was 39,600 L/mol/cm.

5.1.43 p-Chloro-substituted cinnamon-hydroxamic acids

Agrawal, Y. K. et al. [1] developed a selective and sensitive spectrophotometric method for the determination of trace amounts of titanium using N-m-tolyl-p-chlorocinnamohydroxamic acid. It was reported that a yellow extract in chloroform absorbed at 415 nm. The molar absorptivity was 1.4×10^4 l/mol/cm, and Ti (iv) can be determined in the presence of V, Mo, Nb, Ta, Zr, Ce, and Fe etc.

A summary of analytical reagents used for spectrophotometric determination of titanium (IV) is described in Table 5.1.

Table 5.1: Analytical reagents used for spectrophotometric determination of titanium (IV).

S. No.	Name of the reagent	λ_{max} (nm pH/solvent)	Molar absorptivity (l/mol/cm)	Sandell's sensitivity (ng/cm^2)	Limit ofdetection
1.	Ascorbic acid	360			
		3.5–6.0	–	–	–
2.	Methylene Blue and ascorbic acid	665	–	–	–
3.	Hydrogen peroxide and xylenol orange	535	–	–	–
4.	Diantipyrinyl methane	430	–	–	–
5.	Salicylic acid	410	2.5×10^3	–	–
		H_2SO_4			
6.	Nitric acid and H_2O_2	410	–	–	–
7.	Thiocyanate	420	$6–7 \times 10^4$	–	–
8.	N-(2-hydroxybenzylidene)-3-oxo butanehydrazide	500	1.68×10^4	.00285	0.385 µg/mL
9.	Tetraphenylarsonium	420	7.5×10^4	–	–
10.	Alizarin and fluoride	513	7.0×10^4	–	–
11.	p-Hydroxybenzoyl-hydrazone (RAPHBH) & Triton X-100	450	3.1×10^4	1.6	
12.	2-4-dihydroxybenzaldehyde	430	1.35×10^4	10049	–
		500			
		1.5			
13.	Chromotropic acid and oxalate	440	–	0.6	–
14.	2,3-dihydroxy-naphthalene	4–9	3.2×10^4	–	–
15.	2,3-dihydroxynaphthalene (H_2ND)	375	3.2×10^4	.001	–
		4–9			
16.	Haematoxylin and cetylammonium bromide	–	3.5×10^4	–	–
17.	N-phenylauro-hydroxamic acid and phenylfluorone	540	2.33×10^5	–	–
		9–10 M HCl	1.23×10^5		
		560			

Table 5.1 (continued)

S. No.	Name of the reagent	λ_{max} (nm pH/solvent)	Molar absorptivity (l/mol/cm)	Sandell's sensitivity (ng/cm^2)	Limit ofdetection
18.	N-1-hydroxy-Ni,N2 diphenylbenzamide and thiocyanate	400 .05–0.15 M HCl	20000	–	–
19.	Tannin and thioglycollic acid	400 4.0	–	–	–
20.	Diantipyrylmethane with ascorbic acid	390 6.0			0.1 µg/L
21.	Tiron	5.0	–	–	–
22.	2,4-dihydroxy-acetophenone benzoyl hydrazone	371	3.5×10^4	.0035	.0146 ppm
23.	Chromo tropic acid disodium salt and hexaacetate calyx (6) arene	470 3.0	–	–	–
24.	5-bromo-2-hydroxy-3-methoxybenzaldehyde-p-hydroxy-benzoic hydrazone	390 4.0	1.44×10^4	.0033	–
25.	Diantipyrilmethane	390	1.06×10^4	–	–
26.	2,3-dihydroxy-naphthalene	390	3.7×10^4	–	–
27.	o-Carboxyphenyl-fluorone in cationic miceller media	530 Strongly acidic	2.24×10^5	–	–
28.	4-hydroxyderricin or xanthoangelol	433	–	–	–
29.	Thiocyanate and cetyltrimethyl ammonium bromide	421	$1.1–1.0 \times 10^5$	0.47 ng/cm^2	–
30.	Diantipyrylmethane	400	–	–	–
31.	o-Nitrophenyl fluorone and H_2O_2	550 0.016 M H_2SO_4	–	–	.064
32.	Chromo tropic acid	420 Acidic medium	–	–	0.7 mg/L

Table 5.1 (continued)

S. No.	Name of the reagent	λ_{max} (nm pH/solvent)	Molar absorptivity (l/mol/cm)	Sandell's sensitivity (ng/cm²)	Limit ofdetection
33.	Water extract of slippery elm leaf	415	0.68×10^4	–	.0131 µg/L
34.	4-(2-thiazolyazo)-1,2,3-trihydroxybenzene	560 2.0–5.0	4.07×10^4	–	–
35.	2-hydroxy-3-methoxyphenyl fluorine (HMPF)	536 H_2SO_4	1.9×10^5	–	–
36.	Bromopyrogallol red	–	4×10^6	–	0.2 ng/cm²
37.	4-nitrocatechol and tetrazolium salts	430 440	2.38×10^4 4.40×10^4	–	–
38.	Thiocyanate	–	1.9×10^5	–	–
39.	3,4-dihydroxybenzoic acid	380 4.0–5.0	–	–	–
40.	5-chlorosalicylic acid	355 2.5–3.0	1.9×10^4	1.25×10^{-3}	1.2 ng/mL
41.	3-(o-Carboxyphenyl)1-o-tolyltriazene-N-oxide	410 1–3.5	5.74×10^3	8.3	–
42.	2-pyrilidene-iminobenzohydroxamic acid (2-PIBHA)	375	39600	–	–
43.	p-Chlorosubstituted cinnamohydroxamic acid	415	–	–	–

Bibliography

[1] Agrawal, Y. K., Jain, R. K.: Solvent extraction and spectrophotometric determination of titanium with p-chlorosubstituted cinnamohydroxamic acids. *Bull. Soc. Chim. Belg.* 90(5), 437–444 (1981)

[2] Babaiah, O., Kesava Rao, C., Sreenivasulu Reddy, T., Krishna Reddy, V.: Rapid, selective, direct and derivative spectrophotometric determination of titanium with 2,4-dihydroxygenzaldehyde isonicotinoyl hydrazone. *Talanta* 43(4), 551–558 (1996)

[3] Banerjee, N. M., Roy, B. C., Lahiri, S.: 3-(o-Carboxyphenyl-1-o-tolyltriazene-N-oxide. A new triagene derivative for the spectrophotometric determination of titanium (IV). *Indian J. Chem, Sect. A, Inorg. Phys. Theor. Anal.* 26(1), 538 (1987)

[4] Banerjee, S.: Spectrophotometric determination of titanium with tannin and thioglycollic acid and its application to titanium-treated steels and ferrous and non-ferrous alloys. *Talanta* 33(4), 360–362 (1986)

[5] Capitan-Vallvey, L. F., Valencia, M. C., de Orbe, I.: Determination of titanium in sea water by ion-exchange spectrophotometry. *Microchem. J.* 40(2), 166–174 (1989)

[6] Chang, S., Tang, J., Li, Z., Pan, J.: Study on chromogenic reaction of 2-hydroxy-3-methoxyphenylflurone with titanium and its application. *Metall. Anal.* 25(2), 10–12 (2005)

[7] Devireddy, M., Saritha, B., Giri, A., Reddy, T. S.: Direct spectrophotometric determination of titanium (IV) with 5-bromo-2-hydroxy-3-methoxybenzaldehyde-p-hydroxybenzoic hydrazone. *J. Chem. Pharm. Res.* 6(6), 1145–1150 (2014)

[8] Du, X., Xu, Y., Qin, L., Lu, X., Liu, Q., Bai, Y.: Simple and rapid spectrophotometric determination of titanium on etched aluminium foils. *Am. J. Anal. Chem.* 5(3), 149–156 (2014)

[9] Ferreira, S. L. C., de Araujo, N. M. L., Matos, R. C., Costa, S.: Fast spectrophotometric determination of titanium in rocks with 3,4-dihydroxybenzoic acid. *Mikrochim. Acta* 122(1–2), 95–99 (1996)

[10] Gao, L.: Spectrophotometric determination of titanium in aluminium alloys with diantipyrylmethane. *Metall. Anal.* 28(7), 56–58 (2008)

[11] Gunawardhana, H. D.: Extraction and spectrophotometric determination of titanium (IV) with N-phenyllaurohydroxamic acid and phenylfluorone. *Analyst* 108, 952–958 (1983)

[12] Harada, A., Ito, S.: Spectrophotometric determination of titanium (IV) with calcones contained in a Japanese herb “Ashitaba”. *Jpn. Anal.* 58(1), 681–686 (2009)

[13] Hines, E., Boltz, D. F.: Spectrophotometric determination of titanium with ascorbic acid. *Anal. Chem.* 24(6), 947–948 (1952)

[14] Hoshino, M., Kamino, S., Mitani, S., Asano, M., Yamaguchi, T., Fujita, Y.: Spectrophotometric determination of titanium with o-carboxyphenylfluorone in cationic micellar media, and its equilibrium and kinetic studies. *Talanta* 85(5), 2339–2343 (2011)

[15] Huang, X., Zhang, W., Song, S., Wang, L.: Kinetic spectrophotometric determination of titanium (IV) at trace level with bromopyrogallol red in ionic/non-ionic surfactant mixed micellar media. *J. Indian Chem. Soc.* 79(1), 65–67 (2002)

[16] Hultquist, A. E.: Spectrophotometric determination of Titanium using salicylic acid in H_2SO_4 medium. *Anal. Chem.* 36(1), 149–151 (1964)

[17] Jain, A., Prakash, O. K.: Spectrophotometric determination of titanium (IV) with 2,4-dihydroxyacetophenone benzoylhydrazone. *J. Indian Chem. Soc.* 92(3), 399–405 (2015)

[18] Khan, K., Masood, H., Khan, T. A.: Spectrophotometric determination of titanium and iron-application to Cherat Volcanic ash N. W. F. P. *Pak. J. Sci. Ind. Res.* 41(4), 217–220 (1998)

[19] Kika, F. S., Themelis, D. G.: Selective stopped-flow sequential injection method for the spectrophotometric determination of titanium in dental implant and natural Moroccan phosphate rock. *Talanta* 71(3), 1405–1410 (2007)

[20] Kostova, S., Dimihov, A., Alexandrov, A.: Complexation of titanium (IV) with 4-nitrocatechol and tetrazolium salts tetrazolviolet and iodonitrotetraazolium chloride. *Chem. Pap.* 54(2), 66–71 (2000)

[21] Leong, C. L.: Spectrophotometric determination of titanium (IV) with haematoxylin and cetyltrimethyl-ammonium bromide. *Analyst* 102, 293–297 (1977)

[22] Lopez Nuñez, R., Callejon Mochon, M., Guiraum Perez, A.: Extraction and spectrophotometric determination of titanium (IV) with alizarin and fluoride. *Talanta* 33(7), 587–591 (1986)

[23] Ma, G.-Q., Zou, M., Huo, H. Y.: Spectrophotometric determination of titanium in high Ti-bearing blast furnace slag with diantipyrylmethane. *Metall. Anal.* 33(7), 77–80 (2013)

[24] Ma, W.-X., Xue, Y., Fu, Y., Qiu, X., Xu, X.: Spectrophotometric determination of titanium in steel with 4-(2-thiazoylazo)-1,2,3-trihydroxybenzene (TATB). *Metall. Anal.* 27(10), 72–74 (2007)

[25] Moharir, A. V., Sarma, V. A. K., Krishnamurti, G. S. R.: Spectrophotometric determination of titanium with tiron. *Microchem. J.* 17(2), 167–172 (1972)

[26] Mondal, R. K., Tarafdar, P. K.: Extractive spectrophotometric determination of titanium in silicate rocks, soils and columbite–tantalite minerals. *Mikrochim. Acta* 148, 327–333 (2004)

[27] Monji, A. B., Ahmadi, S. J., Zolfonoun, E.: Novel method for selective spectrophotometric determination of titanium (IV) with water extract of slippery elm leaf as a new reagent. *Anal. Lett.* 40(2), 2384–2390 (2007)

[28] Mousavi, H. Z., Pourreza, N.: Catalytic spectrophotometric determination of titanium (IV) using Methylene Blues ascorbic acid redox reaction. *Anal. Chem.* 26(12), 1902–1904 (1954)

[29] Otomo, M.: The spectrophotometric determination of titanium with hydrogen peroxide and xylenol orange. *Bull. Chem. Soc. Jpn.* 36(10), 1341–1346 (1963)

[30] Salinas, F., Jimenez-Arrabal, M., Merras, D. I.: Extraction and spectrophotometric determination of Ti (IV) using 2-pyrilidineiminobenzo hydroxamic acid as chelating agent. *Bull. Soc. Chim. Belg.* 94(10), 713–718 (1985)

[31] Sedaira, H., Idriss, K. A., Abdel-Aziz, M.-S.: 5-chlorosalicylic acid spectrophotometric method for the determination of titanium and simultaneous determination of titanium (IV) oxide and iron (III) oxide content of Portland Cement. *Analyst* 121(8), 1079–1084 (1996)

[32] Srilalitha, V., Prasad, A. R. G., Kumar, K. R., Seshagiri, V., Ravindranath, L. R. K. R.: A new spectrophotometric method for the determination of trace amounts of titanium (IV). *Facta Univ., Phys. Chem. Technol.* 8(1), 15–24 (2010)

[33] Swamidoss, C. M. A., Malkhede, D. D.: Liquid-liquid extraction and spectrophotometric determination of titanium (IV) with hexacetatacalix (6) arene. *Int. J. Appl. Chem.* 10(1), 59–65 (2014)

[34] Tamhina, B., Maul, V., Herak, M. J.: Spectrophotometric determination of titanium as tetraphenylarsonium or tetraphenylphosphonium thiocynate-titanate. *Microchem. J.* 22(3), 275–282 (1977)

[35] Tarafdar, P. K., Durani, S., Saran, Q., Ramaiah, G. V.: Rapid spectrophotometric method (aqueous and extractive) for the determination of titanium in silicate rocks. *Talanta* 41(8), 1345–1351 (1994)

[36] Tarafdar, P. K., Thakur, R.: Micelle mediated extraction of titanium and its ultra-trace determination in silicate rocks. *Talanta* 75(1), 326–331 (2008)

[37] Thakur, R., Jha, R. R., Tarafdar, P. K.: Extraction of titanium and molybdenum with 2,3-dihydroxy naphthalene and their sequential determination in silicate rocks, ores, minerals, concentrates and steel samples by molecular absorption spectroscopy. *J. Indian Chem. Soc.* 90(11), 1975–1982 (2013)

[38] Varghese, A., Khadar, A. M. A., Kalluraya, B.: Simultaneous determination of titanium and molybdenum in steel using derivative spectrophotometry in neutral micellar medium. *Spectrochim. Acta, Part A, Mol. Biomol. Spectrosc.* 64(2), 383–390 (2006)

[39] Vojkovic, V., Allegretti Živĉiĉ, V., Druškoviĉ, V.: Spectrophotometric determination of titanium (IV) by extraction of its thiocyanate complex with cationic surfactants. *Spectrosc. Lett.* 37(4), 401–420 (2006)

[40] Wood, D. F., Jones, J. T.: Spectrophotometric determination of small amounts of titanium in zirconium, zirconium alloys, hafnium, niobium and other metals with diantipyrinylmeth. *Anal. Chim. Acta* 42(2), 215–224 (1969)

[41] Yigzaw, Y., Chandravanshi, B. S.: Extraction and spectrophotometric determination of titanium (IV) with N1-hydroxy-N1,N2 diphenylbenzamidine and thiocyanate. *Mikrochim. Acta* 124, 81–87 (1996)

[42] Zeng, F.-Y., Li, S.-X., Cheng, L.-Q.: Simple and rapid spectrophotometric determination of trace titanium (IV) enriched by nanometer size zirconium dioxide in natural water. *J. Hazard. Mater.* 172(2–3), 618–622 (2009)

[43] Zhou, Z.-R., Zhang, L.-Z., Wu, F.-H., Zhang, J.-W., Su, Z.-W., Zhang, Y.-L.: Kinetic spectrophotometric determination of titanium based on its oxidation effect on discoloring reactions of o-nitrophenylfluorone and hydrogen peroxide. *Metall. Anal.* 28(7), 65–68 (2008)

Index

https://doi.org/10.1515/9783111626833-006